Paving Our Ways

Paving Our Ways
A History of the World's Roads and Pavements

Maxwell Lay
John Metcalf
Kieran Sharp

CRC Press
Taylor & Francis Group
Boca Raton London New York

CRC Press is an imprint of the
Taylor & Francis Group, an **informa** business

First edition published 2021
by CRC Press
2 Park Square, Milton Park, Abingdon, Oxon, OX14 4RN

and by CRC Press
6000 Broken Sound Parkway NW, Suite 300, Boca Raton, FL 33487-2742

British Library Cataloguing-in-Publication Data
A catalogue record for this book is available from the British Library

Library of Congress Cataloging-in-Publication Data
Names: Lay, M. G. (Maxwell G.), author. | Metcalf, J. B., author. | Sharp,
K. G. (Kieran G.), author.
Title: Paving our ways : a history of the world's roads and pavements / Max Lay,
John Metcalf, Kieran Sharp.
Description: First edition. | Boca Raton : CRC Press, 2021. | Includes
bibliographical references and index.
Identifiers: LCCN 2020031512 (print) | LCCN 2020031513 (ebook) |
ISBN 9780367520786 (paperback) | ISBN 9780367520809 (hardback) |
ISBN 9781003056300 (ebook) | ISBN 9781000228465 (epub) |
ISBN 9781000228403 (mobi) | ISBN 9781000228342 (adobe pdf)
Subjects: LCSH: Pavements—History. | Roads—History.
Classification: LCC TE250 .L39 2021 (print) | LCC TE250 (ebook) |
DDC 625.809—dc23
LC record available at https://lccn.loc.gov/2020031512
LC ebook record available at https://lccn.loc.gov/2020031513

ISBN: 978-0-367-52080-9 (hbk)
ISBN: 978-0-367-52078-6 (pbk)
ISBN: 978-1-003-05630-0 (ebk)

Typeset in Sabon
by codeMantra

Advice from Leunig[1]

and

A pavement charter

First let the wayes be regularly brought

To artificial form, and truly wrought;

So that we can suppose them firmly mended,

And in all parts the work well ended,

That not a stone's amiss; but all compleat

All lying smooth, round firm, and wondrous neat.

Thomas Mace, Trinity College, Cambridge 1675[2]

Contents

Preface: What is a road pavement?

Travel has always played an important role in the functioning of the human species. After our ancestors descended from the trees, the character of that role changed as human existence now relied on journeys across the ground. Frequent essential journeys to one particular location, such as a waterhole, led to the incidental creation of tracks and paths. Some paths later became wide enough to be called roads and some of these paths and roads came to have surfaces applied to them. The purpose of these surfaces – now called *pavements* – was to make travel easier or safer or more impressive. Our roads were the silver threads that tied our communities together over both distance and time, and the pavements were the fabric that turned those threads into a useful reality.

This all-pervading role of pavements might be reason enough to write a book about their world history over the four millennia or so of their existence, particularly as no previous history has been published. However, another potent reason is that the construction and subsequent use of pavements reflect the real world in which they existed, presenting their strengths, foibles and weaknesses in an interesting new light. This history will also demonstrate the shaky and stumbling nature of human progress over the millennia.

The book is written for interested laypeople and does not require any prior knowledge of pavement technology or of road engineering. To assist such a readership, two chapters provide optional detours into some of the more arcane mysteries that surround paving materials.

Ground travel was once an occasion in itself. Today, travel by road is often considered as a peripheral event and may only have subconscious impacts on the traveller. Most of us recognise the distance that we must travel as a necessary and unavoidable dimension but, because it is much longer than human dimensions, we often measure our journeys in units of time rather than in kilometres and we rarely realise that the pavements we use also have dimensions of width and thickness and surface texture.

Pavements are sufficiently wide to accommodate vehicles such as cars and trucks. Their total width is usually thought of in terms of the number of lanes, rather than metres. The number of vehicles expected to use the road will determine the number of lanes needed and hence the required pavement width. The broader surface geometry of the pavement – such as its curves and gradients – is a function of the road and is largely determined by the terrain and the desired speed of travel. It is beyond the scope of this book but has been well covered elsewhere.[3] One exception to this exclusion is the common feature in many parts of the world where tracks follow ridge lines. Ridges are commonly stony rather than soft. Such ridgeways also avoid the swampy land that frequently occurs in valleys.[4]

Travellers are often concerned with road curves and gradients and the quality and comfort of the ride, but they rarely, if ever, consider the type and thickness of the pavement on which they travel. Indeed, they often think poorly of road pavements and of those who provide and maintain them. Even pavement engineers can be critical of their colleagues. George Deacon was a well-known 19th century British pavement engineer and a colleague of Lord Kelvin. The opening sentence in his major 1879 paper on street pavements read: "As a branch of Civil Engineering street paving has never occupied a very high place."[5]

The cost of the construction and maintenance of a pavement will depend on its length, local "engineering" conditions, regional statutory requirements and the amount of traffic the pavement will be required to carry. Heavily used pavements and pavements in prosperous localities are usually readily funded by taxes on users and adjacent land owners. At the other extreme, pavements connecting poor remote areas must usually be funded from external sources via some form of community service obligation.

Pavements are thicker than most travellers would imagine. For modern pavements, the thickness to be used is determined by the magnitude and number of the loads to be carried and these are essentially the wheel loads of heavy commercial vehicles. Cars will usually determine the width of a pavement, but trucks will always determine its thickness. Attempts to minimise pavement thickness and thus the cost of the road can give rise to surface defects such as roughness, rutting and cracking of the pavement, which will directly affect ride quality and safety and increase the pavement's subsequent maintenance costs. All pavements deteriorate over time and need routine maintenance, but the process is hastened not only by the passage of heavy wheel loads but also by the effects of climate, specifically water entry and immersion, and by the temptation for various providers of utility services to dig poorly backfilled trenches in any pavement in good condition.

The first road pavements were often built for ceremonial purposes, sometimes to allow the king's remains to be transported in state! Later they were piecemeal attempts to ensure trafficability, first for human and animal foot traffic, then for human- or animal-drawn vehicles. A major change occurred with the Romans, who are rightly famous for their extensive network of essentially military roads built to allow the control and taxation of their Empire, and who consequently did build thick pavements. It wasn't until the much later widespread use of routes predominantly servicing local horse and cart traffic that serious attention began to be paid to the sorts of thickness necessary and the types of materials needed.

Thus, in concentrating the discussion in this book on road pavements, we will assume that others have estimated the magnitude and type of traffic that will use the road and that this has led to decisions about the form of the pavement and to decisions by others about the pavement's width and its overall geometry and alignment – particularly its curves and slopes. We will focus in this book on how, over history, a pavement has been selected, constructed and maintained to carry that estimated traffic over the required alignment. Clearly, the key decision factors which we will focus on will be the type and thickness of the materials used and how they were placed and maintained in position to form the operating pavement and, of course, on how the underlying ground – technically called the subgrade – is able to support the construction of the pavement and its subsequent service under traffic.

When a bridge fails, it clearly can no longer perform its intended function. However, when a pavement fails by such calamities as being washed away, bogging vehicles or causing crashes on ice-covered surfaces, the failures can be seen as short term and readily fixable. In most cases, its performance measures will still be satisfied to some degree or other, as the ways that the pavement performs and resists any degradation of its performance are both matters of degree and of scale rather than of catastrophic proportions.

So, given all this, how does a user or an interested observer then judge whether a pavement is performing satisfactorily? In 1843, the City of Philadelphia asked its renowned Franklin Institute to advise it of *the best modes of paving highways.* Later in the same year, the Institute reported[6] to Council that *A good pavement ought to combine stability and moderate smoothness of surface, with facility of removal and replacement, and be as free as possible from noise and dust. Facility of removal and replacement* might seem a strange objective, but the good burghers of Philadelphia were concerned that some of the proposed new pavements might make it difficult to provide, access or repair the pipes and services that they were increasingly placing beneath the city's pavements. Their concern was real as experience elsewhere indicated that often 50% of a street pavement would have been removed and replaced during its lifetime for reasons quite unrelated to its prime purpose as a means for carrying traffic.[7] When France built a very sophisticated pavement testing machine in the 1980s (Chapter 20), one of its first tasks was to test the performance of potential manhole covers to be used to access services below the pavement surface.

A century later in 1948, when the provision and management of pavements were moving from being subjective and empirical towards being objectively based on technical and economic assessments, two acknowledged experts from the California Division of Highways proposed the following simple performance measures. With some enhancements of the Franklin view, they argued[8] that a pavement should be:

- capable of carrying the intended loads,
- sufficiently smooth to minimise the cost of motion,
- skid resistant, to improve safety and aid turning movements,
- economical to build, operate and *maintain*, and
- sufficiently durable to justify the investment.

With the authors' addition of *maintain*, these measures stand the test of time quite well, although today one might expand "relatively smooth" to retain Franklin's low noise generation and include manageable light reflectance. Recall these five easily understood needs and also how pavement performance might be assessed as we discuss the history and development of pavements over the following chapters.

In recent times, pavements have also been widely used at aerodromes and container terminals. The technologies used are very similar to those used for road pavements although the wheel loads are usually much larger. Thus, with but a few passing references, we will leave their story for others to tell.

Authors

Dr Maxwell Lay is an engineering consultant known for his international contributions to road engineering and his acclaimed international road histories. He is a member of the Order of Australia, former Executive Director of the Australian Road Research Board, and past President of the Royal Automobile Club of Victoria and the Australian Automobile Association. He has been awarded the Moisseif Medal of the American Society of Civil Engineers, the Peter Nicol Russell, Warren and Transport Medals of the Institution of Engineers Australia, and the Gold Medal of Roads Australia. He is the author of *Handbook of Road Technology* (CRC Press, 2019).

Dr John Metcalf has worked in pavement materials and engineering at the British Road Research Laboratory, Queen's University in Canada, and the Queensland Main Roads Department. He was Deputy Director of the Australian Road Research Board and later was appointed Chaired Professor of Civil Engineering at Louisiana State University. He was a Fellow of the (UK) Institution of Civil Engineers and the Geological Society of London and a member of the American Society of Civil Engineers. He is an Honorary Fellow of the Institution of Engineers Australia.

Kieran Sharp worked for the Australian Road Research Board (ARRB) as a researcher in the fields of pavements, roads, and transport, and as technical editor, for over 40 years. He was the first member of ARRB to receive an Austroads Achievement Award. He is the Chair of the Technical Committee of the Road Engineering Association of Asia and Australia, a member of the Governing Council, and an Honorary Member. He was awarded the Roads Australia Award for Technical Excellence in 2013.

Chapter 1

How pavements are affected by traffic and weather

Life on our planet has always involved movement, with its various inhabitants seeking food and shelter and partners. On land, many well-used travel routes developed preferentially as they led to common destinations such as fresh water or used convenient ways such as strips of firm ground between swamps or gaps in mountain ranges. *Firm* is an important term in this book and will imply that, under any likely combinations of temperature and moisture, the layer in question can withstand any realistic construction processes and any traffic-induced forces applied to its upper surface, without limiting its ability to perform all of its intended functions. A very early definition of "firm" was given in an 80 AD description of Roman road construction: *The ground must not give way nor must the bedrock or base be at all unreliable when the paving stones are trodden.*[9]

Commonly used routes utilising firm ground often became pronounced paths across the countryside, with hindering vegetation pushed aside and surfaces trampled by the foot passage of their many users. The resulting ways were usually far from ideal, particularly in terms of their location, gradient and surface condition. Thus, as human societies developed, there were increasing pressures to improve the relevance and usefulness of many ways. The demand heightened as humans learnt how to use animals such as oxen and horses to haul loads. The haulage devices were initially sleds and primitive wheeled vehicles. The subsequent widespread availability of useful wheeled vehicles, of large marching armies and of herds of market-bound animals placed significant demands on the surface condition of strategically located routes.

The purpose of this book is to describe how these demands have been met over the ages by improvements to the *in situ* surfaces and, more commonly, by providing *imported* pavements to enhance those surfaces. It aims to provide a descriptive history of pavements for a general readership, and so it would be wise to begin by defining what a pavement is. The pavement is the upper portion of a road or path. It is placed on top of the natural ground which may have been prepared for the pavement by clearing, excavating or filling. The pavement thus provides a suitable surface for travellers and avoids the underlying ground being damaged by those travellers or by the weather.

The word *pavement* comes from the Latin *pavimentum* which means a rammed or beaten floor.

A broad technical definition of a pavement is: [10]

> A pavement is comprised of a number of horizontal courses, each comprised of a different set of materials. This allows the most appropriate materials to be used for the varying conditions that exist throughout the depth of the pavement structure. These courses are often subdivided into horizontal layers which are portions of a course that can be placed and compacted as single entity.

Pavements are not always needed. In dry climates, traffic could operate over many in situ surfaces without the specific provision of a pavement. Such routes could be destroyed by wind, water or the passage of users. The presence of clay or sound rock could help secure such surfaces, whereas wind could readily remove the surface or cover it with sand dunes. However, the problems caused by water are usually much more substantial and will underlay much of our subsequent discussion.

Given this focus, the book will ignore tasks associated with the realignment of routes, the widening of the right-of-way, the construction of cuttings and embankments to improve gradients, and the building of bridges and fords to cross waterways. These are well discussed elsewhere.[11] The crossing of swampy ground, however, does fall within the ambit of this book and is the subject of Chapter 5.

The routes being discussed will all fit within the common perception[12] of a road, and the term *road* will henceforth refer to travelled ways in general. The operational requirements of a road's pavement are that:

1. It contains no unexpected obstacles that would require a vehicle to manoeuvre laterally during its longitudinal passage along the road.
2. Its surface is smooth enough to provide an acceptable level of "comfort" to occupants and cargo using the road and not to generate excessive noise or raise a vehicle's rolling resistance, which typically accounts for about a quarter of a vehicle's energy consumption.
3. Its surface texture is sufficiently rough to allow a vehicle to accelerate, proceed and brake in acceptable and predictable manners and to safely negotiate curves in the road.
4. It is sufficiently strong and stiff to carry many vertical traffic loads without suffering unexpected damage.

There are also three management requirements:

1. It can be built, cleaned and routinely maintained at an affordable cost.
2. It does not require unexpected maintenance expenditure due to wear and tear from expected usage, and

3. It can meet all the above requirements without being compromised by expected external factors such as weather and water.

These seven requirements will all be discussed in later chapters.

A pavement meeting these requirements has three basic components. There is a *surfacing* which directly carries and services the traffic and protects the remainder of the pavement from the weather. The vertical location of the surface is governed by the required alignment of the road. Next, there is a *basecourse* (or roadbase) which provides the strength and stiffness needed to carry the traffic wheel loads for the intended life of the pavement. It may incorporate the surfacing function. It will commonly be constructed from imported material and may be comprised of a number of horizontal layers. The basecourse sits on a prepared surface known as the *grade* or *formation level*. The underlying material below the grade surface is called the *subgrade*. It is also sometimes called the subsoil or road foundation, but in this book, we will use *subgrade*.

The subgrade is thus the third component of a pavement and supports the construction of the basecourse and its subsequent operation under traffic. Subgrades occur as a result of three quite different circumstances. There might be pre-existing firm ground, some of which would need to be removed as it was above the grade level. The remaining parts of the firm ground become the subgrade. Or some of the ground below the grade surface might be incapable of supporting the pavement and its future traffic. It would need to be removed and replaced with an adequate thickness of suitable imported material, sometimes called a *capping* layer. Finally, if the upper surface of the firm ground is below the required grade level, it would need to be covered by an imported topping layer to achieve the required level. In the last two cases, the subgrade is the remaining underlying ground plus the capping or topping layer.

Capping layers are also used to protect the subgrade from water entry or when the subgrade is found to be too weak to adequately support construction of the operating pavement. Such capping layers are sometimes called *sub-bases*.

Subgrades may be improved by techniques such as rolling, draining and stabilisation – as will be discussed later in the book. Subgrades may be damaged by high loads or by water. Hence, there are two additional structural requirements for a pavement:

1. It distributes the loads applied to it by traffic so that the resulting loads do not damage the subgrade, and
2. It does not permit damaging water to enter the subgrade.

Thus, a pavement is much more than a road surface as it also provides a road with strength, stiffness and durability. Durability is an important

concept as, in any realistic scenario, it is not practical to build and operate a pavement that will last without regular maintenance. Ideally pavement managers should be optimising the distribution of the efforts that
they devote to the construction and subsequent long-term maintenance of a
pavement.[13] This task is complicated by the difficulties of predicting future
traffic and local climate.

While vehicle usage is an obvious influence on pavement performance, a
commonly under-appreciated fact is that even unused pavements will deteriorate over time. Chapter 3 discusses the elaborate "Military Roads" built
by General Wade in Scotland during the 18th century. His expensive roads
changed from occasional use to neglect and then to decay over a quarter
of a century as essential *maintenance work was carried out with waning
enthusiasm and decreasing efficiency.*[14]

Traffic and water are the two major damaging influences on the life of a
pavement. Traffic damage is typically rutting and wear and tear due to the
usage of the pavement and is discussed in more detail in Chapter 19. Water
destroys pavements in many ways, and it will be helpful to outline these
before proceeding further. Water damages pavements by:

- Washing material away, as illustrated in the Preface.
- Reducing the strength and stiffness of in situ materials, particularly
  clays and silts. Sometimes even a small amount of water can cause
  major deleterious changes. However, most paving materials can
  accommodate some water without serious consequences, and many
  have an *optimum moisture content* (see Chapter 4).
- Destroying surfaces by uplift pressures which occur when water fills
  any internal air voids and pores in a pavement. Any loading pressure
  on the surface of the water-logged pavement is then carried throughout
  the pavement by the water acting as an hydraulic system. The resulting
  hydraulic pressures operate in three dimensions and will push aside
  any weaknesses in the pavement structure. This process is sometimes
  called *pumping*, and the pressures are called *pore pressures*. Probably
  the most common example is in pothole formation where the surface of the pavement – designed primarily for downwards pressure –
  cannot resist any upwards hydraulic pressure and breaks free from the
  underlying pavement. Pumping also removes materials from the pavement, and pavement cracks and outlets discharging water are often
  stained with fine materials pumped from the pavement.[15]
- Causing damage by freezing and thawing and *frost heave*.[16] This is a
  major problem in regions subjected to long periods of snow coverage
  and with permeable soils where – even though the pavement surface
  might be impermeable – water can be drawn upwards into a permeable pavement from the subgrade via capillary action. The heaving is
  a consequence of water expanding by about 10% when it freezes and

becomes ice. This causes cracking, allowing more water to enter and more subsequent heaving. Then, when the pavement thaws the ice melts leaving voids which lower the bearing capacity of the pavement and the subgrade. This process can cause major rutting, often called *spring break-up*.
- Promoting root growth and associated expansive pressures.
- Dramatically changing the volume of clays.
- Changing the physical state of exposed rocks such as "green" basalts.[17]

The eight key means for alleviating the pavement damage caused by water are to:

1. Use water-insensitive paving materials.
2. Avoid building the pavement with materials with too high a moisture content.
3. Build measures to control and reduce the level of free water (the water-table) in land adjoining the road.
4. Prevent water entry by using longitudinal side-drains – attention to such drainage is particularly important where the subgrade is below the surface level of the surrounding ground.
5. Maintain an impervious crack-free pavement surface.
6. Provide easy exit points for water within the pavement such as a porous pavement layer on top of the subgrade linked to free-draining pavement edges – these porous layers are sometimes called *drainage blankets*; the porous layer is usually made using a layer of single-sized broken stones (called open-grading, Chapter 4).
7. Otherwise, minimise air voids and passageways within the pavement, and
8. In cold climates, restrict temperature changes within the pavement and subgrade.

Chapter 4 provides a technical background to all these water-related issues.

In dry areas, roads can be quickly covered by wind-blown sand. In timbered regions, trees and shrubs can quickly reclaim a roadway, and in tropical areas, rampant jungle growth can close a road in a single season. The problem also occurred, but at a slightly slower pace, in Europe where, in 14th-century France, Charles VI issued an ordinance noting that hedges, brambles and trees had grown on many roads in the vicinity of Paris, causing some to be abandoned. He ordered force to be used if necessary to have the inhabitants clear the ways.[18] In the countryside, landowners would destroy any signs of a road's prior existence and incorporate the land into their farms.[19] In densely populated regions, people commandeered road space for building houses, herding cattle, as a butchers' slaughterhouse, dumping offal and other rubbish or removing road materials they found personally useful. The authors have recorded[20] cases of citizens in our home town brazenly stealing paving blocks from existing roads.

It might be appropriate to end this chapter with an instructive story from Aylesbury about 100 km north-west of London. In 1499, a miller needed to repair his mill, so he sent his staff to obtain some useful "ramming clay". The best clay was on the highway where their excavations created a pit that was about 2 m deep, and of similar width and length. It was winter, and the pit soon filled with water. An unsuspecting traveller on horseback fell into the pit, and he and his horse drowned. The miller was charged with murder but was acquitted as it was considered that he had had no malicious intent and as the highway was the best place to find the clay he needed.[21]

Chapter 2

Early pavements in Egypt, Mesopotamia, Crete and Greece

The first paths and roads would have been created by travellers pushing aside or removing any obstacles that were impeding their intended journey. Once a clear route was established, travel needs might have then directed effort towards improving the surface of the way. The work would initially have been minor and incremental, leaving no records for future observers like ourselves.

The oldest existing pavement dates from about 2500 BC when Egypt's Old Kingdom was actively involved in building pyramids and temples. Much of the construction was in the Giza area near Cairo, but the nearest suitable stone had to be obtained from the Gebel Qatrani quarries around El Faiyum, about 70 km to the south-west. A 2.1 m wide road was constructed to enable the desired basaltic stone to be brought to Giza and about 12 km of that road – now known as the Lake Moeris Quarry Road – still exist.[22] It was built using slabs of limestone and sandstone with original dimensions of about 500×500×100 mm.[23] The Egyptians of the time did not have animal-drawn vehicles, and the stones would have been hauled on sleds or logs rolling on the pavement surface. The Greek historian Herodotus, writing in about 450 BC, noted a stone-paved road about a kilometre long and 20 m wide built near the pyramids to carry the limestone blocks on the final stage of their journey. He wrote *For 10 years the people laboured in the construction of the causeway. which was about a kilometre long, 20 m wide and 17 m high and made from hewn stone sculpted with figures.*[24] In about 1170 BC, a stone-paved ceremonial route was built between two temples at the holy city of Bubastis, about 40 km north of Cairo.[25] Herodotus described[26] it as *A stone road about a kilometre long leads to the entrance, leading eastwards toward the marketplace; this road is about 100 m wide, and bordered by trees so tall that they seem to touch the sky.*

The earliest "urban" pavements that have survived are monumental plazas and processional ways. Many of the best-known ceremonial pavements were built by Mesopotamian monarchs in the delta land between the Tigris and Euphrates rivers – the "fertile crescent". These grand projects were built during the four millennia prior to the better-known Roman roads and were ostentatious displays of wealth and power rather than of engineering skills.

The skills used were largely those of the brick-workers and the stonemasons and had been honed in building construction where durable bricks and selected stones were worked into massive self-supporting shapes. Thus, their roads were also made from bricks and stone blocks. Good photographs of existing block paving in Nineveh dating from about 700 BC are available via online sources such as Wikipedia Commons. The ancient masons chipped away at their raw rocks using flints and bronze- and iron-tipped tools, although iron was still a rarity until at least the millennium before the Romans. Good building stone was not common in Mesopotamia. Quarries in adjoining lands producing flint, greenstone and limestone were highly prized and their products brought to Mesopotamia. Layered limestone was also imported as it could be easily split into stone slabs.

Thus, in the absence of suitable stone, early manufactured pavements in Mesopotamia predominantly used bricks made locally from sun-dried mud, burnt bricks made by pressing clay and then heating the brick in an oil-fired kiln and imported slabs split from layered limestone. They were sometimes bonded together using the bitumen that occurred naturally in the region and was widely used for waterproofing and for sealing tanks.[27] In the biblical Old Testament, Genesis 11(3) notes the call *Come, let us make bricks and burn them thoroughly* and tells us that they had *brick for stone and bitumen for mortar*.[28] This and other recent translations use the word *bitumen*, which comes from an earlier attempt to translate *pitch* into Latin as *pitchumen*, which was then further translated into modern languages as *bitumen*. Some translators used terms such as *slime* or *pitch* rather than *bitumen*. The verses also promote the construction of the Tower of Babel, exhorting *Let us make bricks and burn them thoroughly. And they had bricks for stone and bitumen for mortar.* Later, Genesis 14(10) reads *Now the Valley of Siddim was full of bitumen pits*.[29] Siddim is thought to have been at the southern end of the Dead Sea. The ancient Greeks called the Dead Sea "Lake Asphaltites", which leads to the current word *asphalt*. In construction, the bitumen was often mixed with sand to produce a *mastic* which made the material cheaper and less prone to softening when subjected to high temperatures. Bitumen, asphalt and mastic are discussed further in Chapter 11.

Pavements of bricks mortared with bitumen have been found at Khafaje 10 km east and at Hit (or Heet) about 160 km east of Baghdad in Iraq and in Mohenjo Daro in Pakistan and have been dated at about 2000 BC. The bitumen in the joints also seeped into the bricks and over time this produced a very strong bond between bricks.[30] Of course, bricks mortared with bitumen were even more widely used in building construction. Similar brick pavements surfaced with limestone slabs have been found in Assur 300 km north of Baghdad and Babylon about 80 km south of Baghdad and dated at about 700 BC.[31] There is also an unconfirmed suggestion that some Inca roads in Peru also used stones covered with bitumen.[32]

Indeed, given the ready availability of bitumen, of ways of heating it to reduce its viscosity, of sand and of stone chippings from the works of stonemasons, it would seem quite possible that bitumen and asphalt (a mix of bitumen, sand and stone) might have been used in their own right as pavement surfaces in Mesopotamia. However, no records exist to confirm this.

The earliest paved street is thought[33] to have been constructed by the Hittites in Boğazkale in central Turkey in about 1200 BC. The best-known Processional Way was built in about 600 BC by Nebuchadnezzar in Babylon using a technique introduced by his father, Nabopalazzar.[34] The son created an inscription beside the Way which read that his father had initially made the Way *glistening with bitumen and burnt bricks*.[35] It was called *The street on which may no enemy ever tread*. The Way ran north-south and was almost 2 km long, and Nebuchadnezzar had paved it with limestone flagstones about 350 mm thick placed on three horizontal layers of burnt bricks, all mortared together by lime and bitumen. The underlying subgrade appears to have been rammed earth and sand.[36] If sand is well-compacted and restrained from horizontal movement, it can provide an adequate subgrade – a point that has eluded many subsequent non-technical commentators. Indeed, sand has the virtues of being inert and free-draining. The floor of Nebuchadnezzar's temple in Babylon was paved with burnt brick blocks mortared with bitumen, and there are photographs of parts of the floor remaining in about 1920.[37]

The Mesopotamian empires built extensive arterial roads to maintain control over their territories, but often they were little better than rammed in situ materials, although Diodorus writing in about 50 BC suggested that part of the Royal Road between Susa and Ecbatana (in modern Iran) was paved.[38] There was little demand for manufactured roads outside the cities and this demand diminished further as the power of the Mesopotamian empires declined.

The Minoan civilisation in Crete from about 2600 BC to 1100 BC developed relatively sophisticated road-building methods which probably influenced later Etruscan and then Roman road builders.[39] Indeed, it has been observed that they appeared to rely on the pavement as a structure more so than did the Romans whose prime emphasis was on thickness and subsequent load spreading, although the Romans did have to deal with much heavier wheeled traffic.

A fine example built near Knossos in about 1600 BC is shown in Figure 2.1. It appears to have been mainly built for foot traffic. The Minoans placed a layer of 50 mm stone blocks on a firm subgrade using cementitious mortar between the blocks to render the layer watertight. The blocks were then covered by a 50 mm cementitious layer to provide a bed for basalt paving slabs. These slabs were typically 50 mm thick and also mortared. Early in the 20th century, Evans[40] noted that the excavated pavements were still in

Figure 2.1 Portion of existing original Cretean road near Knossos.[42]

"perfect" condition and showed no signs of wheeled traffic. Cementitious materials are discussed further in Chapter 11.

As the Greek empire developed in the Mycenaean period (c1600–1100 BC), administrative demands led to some roads being constructed in hilly and precipitous regions. However, the rocky nature of the ground meant that constructed pavements were not a major issue.[41] Ruts were created in many rocky pavement surfaces by the wheels of various hauled vehicles and some were deliberately made to provide secure passage in mountainous terrain (Figure 2.2)[43] or to maintain orderly passage in towns. The usage in towns can still be seen in the ruins of Pompeii which had been covered by an eruption in AD 79. There is evidence that the inhabitants had used molten iron to protect the worn base of some ruts. There are many good photographs of these ruts at online sites such as Wikipedia Commons.

Figure 2.2 Deliberately rutted Roman road in the hilly Bozbergs near Effingen in Switzerland's Aargau canton. The rut spacing is about 1.2 m and the rut depth about 400 mm.[47]

The ruts did create right-of-way problems when vehicles encountered each other head-on, as in the story of Oedipus meeting and then slaying his oncoming and unrecognised father Laius on the Cleft Way near Delphi.[44] Ancient pre-Roman rutted cart tracks have also been found in Greece, Etruscan Italy, Malta and Austria.[45] In 221 BC, the Qin First Emperor in China mandated a standard wheel spacing to ensure that carts and carriages could use existing road ruts.[46]

Major pressures for improved surfaces for travelled ways came from the introduction of wheeled vehicles hauled by large animals such as oxen.[48] By 2000 BC, wheeled vehicles were common in the Middle East and in parts of south-eastern Europe. In particular, the Celts were building quite sophisticated vehicles, as demonstrated in Dejbjerg vehicle in the National Museum in Copenhagen, and there is evidence[49] of an extensive trading network in valuable items such as amber. Roads dating from 2000 BC have been found in swamps throughout Europe (Chapter 5). However, no major surficial remnants have been found of Celtic roads.

The Etruscans in central Italy were paving the streets of their towns in about 400 BC. Four paved roads from this era have been found in Bologna.[50] Despite earlier claims, there is no physical evidence of Carthaginian roads, and the record depends solely on Servius who when writing in about 400 AD noted, *The Carthaginians are said to be the first that paved the Ways, afterwards the Romans did the same.*[51]

Rather than now follow a chronological path and discuss Roman roads which were technically a quantum leap forward, we will next examine the various simpler pavements that developed around the world.

Chapter 3

Making pavements using local materials and simple equipment

Roman roads and their pavements are justifiably well known as they were superior to any previous facilities and were subsequently unsurpassed for almost two millennia. However, before making a detailed examination of Roman pavement construction in Chapter 6, it is useful to begin by studying the many more primitive techniques that had been in use prior to and during Roman times and which in many places continue to be in use to the present day.

Many roads have evolved gradually over long periods of time. The initial paths and tracks have had their surfaces cleared and smoothed and their alignments eased. In this book, we concentrate on the road structure placed above the natural formation – the pavement – and we saw in Chapter 1 that time and weather can quickly destroy pavements, even in the absence of traffic. Thus, to have an operating pavement requires a commitment of efforts and resources to both provide the pavement and then to maintain it. There are many examples of where the enthusiasm to construct a pavement has not been matched by any subsequent efforts to maintain it. This applies to pavements ranging from cleared surfaces to motorways, and even the change from a cleared track to some form of constructed pavement implies serious forward commitments.

In a prosperous community, pavements will be built because they contribute to the wealth and well-being of that community. However, in many places there will be strong competition for the funds needed to build and maintain pavements. A pavement might only be supplied for urban streets or where a rural road provides essential accessibility or carries significant traffic. For example, in some jurisdictions a road carrying more than 100 powered vehicles a day might be considered for formal construction and maintenance.

The economics of pavement provision are discussed further in Chapter 19 once the role of pavements has been more clearly explained. As a first step, we will now consider the basis for the construction and maintenance of pavements. When making the simplest form of new road, the pavement and the subgrade are one and the same thing, with only the local vegetation and associated top-soil removed as unsuitable for a permanent pavement.

Some weaknesses in the remaining material can be remedied by compacting it to reduce further deformation and minimise water damage. Compaction can occur incidentally via the feet, hooves, and wheels of passing traffic. As the importance of the route increases, compaction can also be achieved deliberately by forceful ramming and rolling of the soil. The word *cause-way* was commonly used to describe such compacted routes although it more specifically applied to a single line of stones placed in a footpath or wheelpath.[52] Major advances in the theory and practice of compaction were made in the 20th century and will be discussed in Chapter 4.

Sands and clays can sometimes make useful low-cost pavements, but they also have major disadvantages. Sands are easily moved aside by wind, water and traffic. Clays become soft and sticky when wet. Many clays below a certain critical moisture content can be good mortars and so mixtures of sand and clay have been effectively used when the pavement shape and drainage have kept the clay reasonably dry. In successful applications, the amount of clay used is just enough to fill the air voids in the sand layer and the mixing water is just enough to make the clay sticky.[53] These natural pavements are commonly very dusty when dry (Chapter 14). Clays are complex materials and are discussed in more detail in Chapter 4.

When a desired route had a firm but uneven surface there were many practical ways of smoothing that surface by removing obstacles and importing sand and stones to fill voids. Apart from the physical labour, this could require some masonry skills and some haulage capacity. But the problem would rarely be insuperable. The Telford guideline[54] for his work in the Scottish Highlands late in the 18th century (see Chapter 8) was that any material above firm rock was to be removed if the rock was within 600 mm of the surface. In recent times, loading tests have been developed to ensure that a subgrade will be adequate – the simplest of these is to take a heavy roller over the subgrade and measure any temporary (elastic) and permanent (plastic) deformations. The measurements are assessed and where necessary any soft spots are rectified.

A slightly more sophisticated approach had developed by the beginning of the 20th century. A flat plate was placed on top of the subgrade and it was then loaded to determine what load could be applied to the subgrade without causing noticeable deformation or distress. This will depend first on the area of the plate and is accounted for dividing the load by the area over which it is applied. This is called *pressure* at the contact surface and *stress* within the material. In Chapter 16, we discuss more details of the specific pressures and stresses beneath a motor vehicle tyre. The unit of pressure is the Pascal (Pa).

If the pressure that a subgrade surface can withstand is less than that which would occur beneath the wheels of passing traffic, then a pavement of sufficient thickness would need to be added to reduce the pressure on the subgrade to the acceptable level. One assumption made was that the pressure would diminish with the cube of the pavement thickness; another was that the pressure dispersed through the pavement at an angle of about 40° to the vertical.[55] The fact that the load-induced stresses do disperse and thus diminish with depth explains why cheaper materials can be used in the lower layers of a pavement. Load dispersion is explored further in Figure 19.2. In Chapter 16, we will also re-encounter the loaded plate test which became the basis of the California Bearing Ratio (CBR) subgrade test developed in the 1940s. Incidentally, the development of the empirical "CBR vs required pavement thickness" design chart buried any assumptions about load dispersion or pavement properties far from sight. Not all subgrade properties are immediately apparent, particularly the effects of water and temperature change. Methods of detecting these developed over time and will be referred to in subsequent pages.

The subgrade was defined earlier as the prepared supporting platform for a pavement. The preparation needs to ensure that its surface is smooth as typical construction processes will result in much of its shape being reflected in the shape of the pavement surface. The process used to achieve the required shape of both the subgrade and the pavement surface is called *grading*, and *graders* (or levellers) were some of the earliest tools used by road-makers and remain in use to the present day. The first graders would have been logs dragged along the pavement surface, usually at an angle so that excess material was pushed aside. The *King Drag* was widely used in America early in the 20th century. It used a 200 mm log split down the middle, and the two halves were then braced about a metre apart. The front half-log was shod with a 5 mm steel plate (Figure 3.1). The logs were later replaced by planks, first of timber and then of iron followed by steel (Figure 3.2). In the late 19th century, equipment was available with the scraping blades held by mechanisms that permitted their operating angles to be adjusted (Figure 3.3). The first such device was probably the "American Champion",

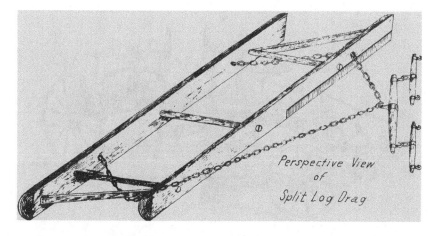

Figure 3.1 The King or Split Log Drag or grader.[56]

Figure 3.2 Grading with a horse-drawn Drag in Victoria, Australia.[57]

patented in 1877 by Samuel Pennock in Pennsylvania. Leaning wheel versions were introduced by Joseph Adams in Indianapolis in 1885. These early devices were often called *road machines*. Self-powered graders were first produced by Richard Russell in Minnesota in 1919. Today, graders can both trim the formation and add pavement materials to the required levels using laser-based controls.

Gravel is a collective term for loose stones, typically smaller than 100 mm, and is at the other end of the spectrum to the sands and clays discussed above. If a low-cost pavement is being built in a region with available gravel, it might well be considered for use in subgrade and pavement construction. However, one common problem with gravels is that they are often found in river beds and glacial drifts where they been smoothed and rounded over many centuries. Such rounded stones are difficult to compact into a coherent and impermeable layer as they are easily pushed aside. This leads to the more common use of broken (or crushed) stones, as will described in following pages. However, a century later rounded stones found widespread use in cement concrete roads (Chapter 17).

Figure 3.3 "American Champion" grader with its adjustable scraping blade. First patented in 1877 by Samuel Pennock in Pennsylvania.[58]

When gravels are used, they are often graded and stockpiled into a range of sizes. Small amounts of clay or fine limestone might also be used as a pseudo-cement. The quantities used from each size-stockpile are selected such that the smaller stones will fit into the interstices in the layer of larger stones. This makes compacting the layer more likely to produce a pavement which has adequate strength and stiffness. However, if only rounded river gravel is available, useful compaction will be unlikely.[59]

Gravel roads came into more prominence in the 19th century as vast territories were colonised in the New World and when road-making machinery became available to the road-makers. Thus, an early 20th-century gravel pavement might be constructed using gravel no larger than 40 mm in size and separately roller-compacted in three layers of 150 mm, 100 mm and 40 mm initial thickness until no further overall or local settlement occurred. The final pavement would be about 200 mm thick.[60]

Over much of human history, most pavements were constructed in fertile areas. In many such cases, otherwise firm soil is quickly rendered impassable by the presence of water and the passage of previous travellers. As their moisture content increases, all soils lose strength and finally turn into fluid mud. Even layers of granular rock particles can lose strength once excess water has filled in the inter-particle spaces and allow any applied surface loads to be carried hydraulically by water pressure (often called *pore pressure*) and not by inter-particle contact. It is therefore of critical importance to control the moisture content of a pavement both during construction and in subsequent service. This is particularly because once water has entered a pavement it is very difficult to remove. Indeed, in a well-built and maintained pavement, the moisture content under the pavement surface will soon reach an *equilibrium moisture content* which will then alter little for the remaining life of the pavement. This is discussed further in Chapter 4. An overall view of the movement of water in a pavement is shown in Figure 3.4.

The water that causes these problems may come from rain falling onto a permeable pavement surface and then passing into and through the pavement to cause damage to both the pavement and the subgrade in the various ways that were summarised in Chapter 1. Other common undesirable situations are for water to be captured in wheel ruts made by the passage of travellers or by the road itself being constructed lower than surrounding land and then acting as a storage pond for any adjacent surficial water flows. Not surprisingly, these poor construction outcomes are often associated with inadequate longitudinal roadside drainage.

A 1562 Act had given English parish road "supervisors" the authority to divert any *water-course or spring of water* into a ditch adjoining the road.[62] The role of the local parish is discussed further in Chapter 7. The first book on roads written in English was produced by Thomas Proctor in 1610.[63] With much common sense, he highlighted the problems created by water which he said led to the *rotting and spoiling of all highways*. Proctor thus emphasised the need for longitudinal side drains, a

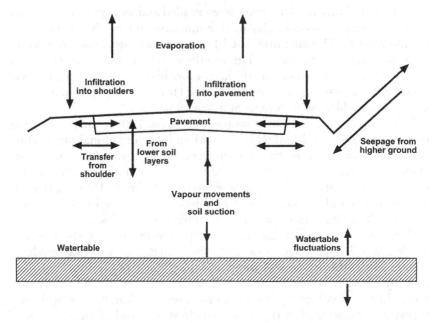

Figure 3.4 Water movement associated with a pavement structure.[61]

raised pavement put in place with *much ramming and treading* and a modest cross-fall at the pavement surface. He suggested that carts be prohibited from using the pavement during the English winter. There is little evidence of such good advice being taken or powers exercised.

The situation became of increased concern as land-based trade became more common in the 17th century. Discussing 1685 in his *History of England* [64] Thomas Macaulay commented:

> On the best lines of communication the ruts were deep, the descents precipitous, and the way often such as it was hardly possible to distinguish, in the dusk, from the unenclosed heath and fen which lay on both sides. ... It was only in fine weather that the whole breadth of the road was available for wheeled vehicles. Often the mud lay deep on the right and the left; and only a narrow track of firm ground rose above the quagmire...coaches stuck fast, until a team of cattle could be procured from some neighbouring farm, to tug them out of the slough......
> The great route[65] through Wales to Holyhead was in such a state that, in 1685, a viceroy, going to Ireland, was five hours in travelling fourteen miles, from Saint Asaph to Conway. Between Conway and Beaumaris he was forced to walk a great part of the way; and his lady was carried in a litter. His coach was, with much difficulty, and by the help of many hands, brought after him entire. In general, carriages were taken to pieces at Conway, and borne, on the shoulders of stout Welsh peasants, to the Menai Straits.

The muddy pavements had some major direct effects on travellers. In 1690, a Chancellor Cooper wrote to his wife attributing the long legs of Sussex women to the constant pulling of their feet out of the mud as they travelled, which he concluded would *tend to strengthen the muscles and lengthen the bones*.[66] A report of circumstances in England in 1737 suggests that it was not uncommon for horses to have to proceed along a road with water up to their bellies.[67] In 1745, the British government was almost overthrown on account of the bad state of British roads.[68] In 1770, Arthur Young commented on ruts in Lancashire pavements: *I actually measured ruts 4 foot deep, and floating with mud only from a wet summer; what, therefore must it be after a winter?*[69] The inconvenience could be permanent as there have been various instances of people drowning in potholes[70] and of horses and oxen abandoned if stuck in deep mud. Things were no better in the New World. A traveller in 18th-century Virginia described its roads as *hopeless seas of mud with archipelagos of stumps*.[71]

An extreme example of the effects of water was the *holloway* where traffic, poor maintenance practices (Chapter 7) and water flows over the centuries have eroded many roads until they are well below the level of the surrounding countryside. On portions of the Silk Road near Samarkand, centuries of use have turned parts of the road into trenches up to 2 m in depth.[72] The naturalist Gilbert White recorded some roads in the region of Selborne in Hampshire in England which were about 5 m below the surrounding fields. In about 1779, he wrote:[73]

> the two rocky hollow lanes, the one to Alton, and the other to the forest, deserve our attention. These roads, ..., are, by the traffic of ages, and the fretting of water, worn down through the first stratum of our freestone, and partly through the second; so that they look more like water-courses than roads; and are bedded with naked rag for furlongs together. In many places they are reduced sixteen or eighteen feet beneath the level of the fields.

In about 1820, a Rev. John Marriot, a Devonshire vicar, celebrated holloways in an eight-verse ballad called The Devonshire Lane which included the following somewhat unclerical verse:[74]

> The banks are so high, to the left hand and right,
> That they shut up the beauties around them from sight;
> And hence, you'll allow, 'tis an inference plain,
> That marriage is much like a Devonshire lane.

The holloways were often quite narrow and few survived beyond the 18th century due to pressures from the land-owners (in England, via the enclosure laws), the increase in cart and wagon traffic and the growth of privately operated turnpikes.[75]

The "let nature take its course" approach to pavements had some equally almost implausible outcomes. For example, this extreme method seemed appropriate to General Wade when he was, at great government expense, attempting to make roads in Scottish Highlands early in the 18th century. Wade was a career soldier who had no road-making skills but had at his disposal ample supplies of military labour and of gunpowder. Writing in 1828 the Chief Inspector of Roads in the Highlands commented:

> The usual practice in the roads constructed by General Wade was to excavate all the earth and turf until gravel appeared, when the road surface was then dressed and formed on it; and the excavation and rubbish being thrown on either side, and the road was left in the form of a ditch.

This would have created a new eroding waterway and the Inspector incidentally noted that it collected snow in the winter.[76] Other accounts suggest that the base of the excavation was first covered with any available large boulders with no attempt to apply masonry techniques to aid fitting of adjacent stones.

Local drains and quagmires in colder climates notoriously collect snow drifts, rendering roads in snow-prone regions impassable. Nevertheless, in 1737 Phillips argued, by selectively observing rocky river beds and coastal sands, that allowing roads to function as occasional rivers, would lead to sound roadways.[77] Phillips was an observer of road-making and -mending, and not a practitioner. A trial of his theory – properly described as *incredible* – at Ware, about 30 km north of London not surprisingly produced disastrous results.[78]

There were some, possibly better, aspects to Wade's work as Haldane (a social historian) also noted – without source – that *where soft ground had to be crossed, the foundations were made with successive layers of stone of diminishing size, the top layer and the surface being gravel.* If not done with great care (as discussed below) such a pavement would soon be pushed aside by wheels and hooves or sink into the soft ground, particularly if the gravels used were rounded rather than angular.

Recognising the value in keeping water away from the pavement, some maintainers piled the paving materials into longitudinal ridges. This was called a *"convex"* road (Figure 3.5). It soon deteriorated (cross-hatch in Figure 3.6) and was maintained by piling more materials onto the centre of the road to produce an almost unusable *"barrell"* road (Figure 3.6) which under traffic soon took its "worst state" form is shown in Figure 3.7. The two practical advantages of the convex road were that: (1) it could be easily constructed by simply dumping materials in the middle of the road and then letting it slump into place, and (2) rain-water would more readily run off the road and out of the pavement and into side drains. This proved a difficult pavement for travellers. However, the convex shape *often approaching the form of a semicircle, was such that vehicles kept entirely to the centre of the road.*[79] Wheeled vehicles straddled the ridge and this concentrated all their damage in one set of wheelpaths, accelerating pavement damage and preventing either of the intended outcomes from occurring. The ridges were quickly flattened so that longitudinally the pavement looked more like a

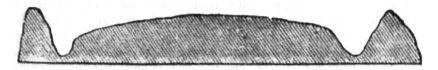

Figure 3.5 Common convex road in 1809.[81]

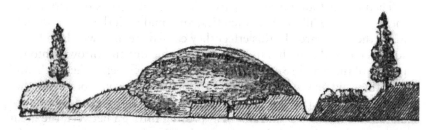

Figure 3.6 Barrell road[82] "Barrell" refers to the shape of the cross-section. The original caption referred to the road in its "second state". The "first state" is the cross-hatched pavement in the lower part of the Figure.

Figure 3.7 Degraded road[83] The original caption refers to the road in its "last and worst state".

half-buried pipe or barrel than a roof. As with the ploughman's road discussed below, any ruts were sometimes then flattened by harrowing. The whole process had little technical merit but did ensure that the longitudinal drains were functional.[80] Convex roads in exaggerated form were sometimes called *roof roads*, or *angular roads* if made in a single cross-slope.

The technique received "official" endorsement in England in 1775 when the Annual Register[84] reported that *An experiment was made, a few days ago, on the Ipswich road, with the plow, contrived to throw up the sides of roads in order to raise them in the middle, and it answered beyond expectation.* Addison[85] has sceptically commented *Whose expectations?* Nevertheless, Law and Clark report that the convex road was the most prevalent form found in Britain in 1809.[86]

The following observations of roads near London in 1737 give a verbal picture of English road-making early in the 18th century:[87]

These ways, and the laying on different sorts of natural gravel, having been found of little advantage, they have, of late years, screened the gravel, so as to lay only the stones upon the roads, most of them being as big as eggs, but this was found to be extremely inconvenient, for the stones beat the horses' feet to pieces, made them so as to make it very hazardous travelling on those loose stones, and very much increased the difficulty of the draught of carriages.

Then it was found necessary to join some finer parts of matter with theses stones to fill up their vacuities, and bind with the stones so as to make the road hard. To this effect they caused the mud which had been squeezed into the ditches in Winter, to be dug out and thrown onto the middle of the road in Summer, about a foot or eighteen inches high, and then they laid on the great stones, in hopes that all would bond together and make the road good. But this did by no means bind so as to make an hard road, for the rain that fell, passing through between the great stones, only moistened the dust and clay that had been thrown on the road before they had been laid on, so as to reduce it to mud, which, being a fluid, could not sustain the stones above it, therefore without binding. The stones would pass through by their greater specific gravity, and change places with that mud; but what hastens this change of place is the pressure of heavy carriages which soon press down the stones and squeeze up the mud.

The next thing to be considered is the manner of laying these large stones upon the road. These stones are generally laid on ten or twelve feet wide, and three or four foot high in the middle of the road, leaving a passage on each side, and this is usually done in the Summer-time, after the dirt and dust has been thrown on, as we have mentioned before. See Figure 2 below.

The carriages and passengers, at first, avoid the stones and go on each side, till those sides, as the rains come on, and the wet season advances, become deep and unpassable, as not having been mended at all, or having been made worse by the dirt and the dust thrown out of the ditches; then necessity obliges the horsemen and carriages to hobble on the stones, which soon sink down to a flat, by reason that the sides, being softened with mud and dirt, yield to their spreading; and likewise what had been dust under the stones becomes dirt, softens daily by the rains, and, as we have said before, the stones sink to the bottom, and the mud rises to the top: so that many hundred loads of stones may be laid on a road for a great width, and to a great height in the Summer, or between Midsummer and Michaelmas (which is generally the time) and perhaps not a stone to be seen by Christmas Day. This appears by Tyburn and Kilburn Roads this year; for the great quantity that was laid on to a great height and width at the latter end of August, or beginning of September, was almost all sunk to the bottom by the end of October, the mud and the dirt being got to the top.

In 1756, London's Mile End Road was described[88] as *a stagnant lake of deep mud from Whitechapel to Stratford* (a distance of about 8 km on today's A11). The situations described above were common in England up to the 19th century.[89] Large stones tipped into poor ground do not stack themselves into neat vertical piles. Instead, due to minimal horizontal resistance they often find it easier to tilt or to move horizontally away from their "intended" location. The belief that tipping large stones into poor ground will eventually produce a firm road foundation has persisted to the present day. In our own professional experience, we have encountered cases where local subsidence of an asphalt road over many decades was treated by regularly placing more asphalt over the subsided areas. The roads continued to subside and over time the many-layered asphalt became some metres thick.

The silliness of returning muddy material from the roadside to the road would have been obvious and anyone who had walked on a gravel pavement would have observed how much better the pavement was when the stone used was sharp-edged rather than smooth, rounded gravel. So how did these appalling English roads occur?

In 1763, one writer[90] described roads in the preceding decades as looking *more like a retreat of wild beasts and reptiles, than the footsteps of man.* Despite the poor outcomes discussed above from simply pushed wayside materials – soil, vegetation, cut branches, rubbish – into the muddy remnants of the pavement, the English Highways Act of 1767 repeated an older 1696 law when in paragraph 27 it stated that it was lawful for *surveyors to carry away the rubbish or refuse stones of any quarry as they shall judge necessary for the amendment of the highways.* It was also lawful to take suitable road-making materials from land adjacent to the road, although any pits had to be filled within a month.[91]

Some communities even used the roadway as a place to dispose of their local rubbish and domestic waste, others quarried the roadway for good soil.

For example, the Droitwich toll road trust in Worcestershire issued the following notice in 1767:[92]

> Persons must not take away manure from the side of the road as it is wanted for filling up the holloways.

Some reversal appears to have occurred as a decade later the Trust issued a further notice:

> Persons who have laid manure upon the road (shall have it) removed within a fortnight.

Using organic materials for pavements was apparently not uncommon and, in 1805, Bramley wrote in his "Road Maker's Guide":[93]

> The first principle apparently impressed by the Almighty Power upon all inert matter is a tendency to become proper for the system of vegetation ...... from this cause it becomes the proper system of the road maker to select such materials for that use as are most distant from the capacity of supporting vegetables.

The view was somewhat out of character as Bramley subsequently made many sensible comments. In France, a 1601 ordinance required solid and liquid rubbish and refuse to be collected from the streets of Paris to depots on the Paris to Meaux road, where it would be used to repair the pavement.[94]

If large pieces of rock occurred in surrounding fields, this could lead to both land-owners and road-menders rolling the boulders into the pavement zone, displacing mud but making the pavement even more hazardous. Writing of Oxfordshire in the 1770s Arthur Young commented[95] *The two great turnpikes which crossed the country ... were repaired in some places with stones as large as could be brought from the quarry.* River gravel was often the only practical source of rock and, as noted earlier, the smooth, rounded nature of these stones make them almost impossible to compact. Obtaining stone by quarrying was rarely an economic proposition. A 1532 English Statute under Henry VIII required London roads to be made using cobblestones brought from the seashore (Kent?). In the 19th century, the development of iron-making produced large quantities of *slag* which provided a useful new source of stone suitable for pavements.[96]

Obviously, most road-making remained quite primitive. Turnpike trustees were often vague in their requirements. The Bedford-Bagshot trustees merely required their Surveyor to keep the turnpike *in good repair.*[97] John Scott, in a 1778 review of turnpike pavement repair practice, felt obliged to suggest that stones used for repair should at least *be spread evenly over the road surface so that vehicles did not come into contact with the road's foundation.*[98] One maintenance practice used to manage rapid pavement deterioration was the *ploughman's road* in which the travelled surface of the pavement was kept relatively horizontal and raised up to 2 m above the surrounding countryside – a geometric practice used in the Roman roads to be discussed below. However, each spring a special plough, drawn by a team of eight or more horses, was then used to turn over the pavement surface rutted and distorted by the winter traffic, throwing the ploughed material towards the centre of the pavement.

A specific and quite indicative example of contemporary road-making was provided in 1806 by the trustees for a toll road between Farnsfield and Mansfield in Nottinghamshire. Its request for tenders to build the road was as follows (*ling* is a type of heather):[99]

> The road to be formed 40 ft wide and to be made what is called a concave or hollow road, 18 ft of the soil or turf to be cut in the middle of such depth as may receive the ling and gravel. [then paraphrased] The ling is to be placed first, then 150 mm of unscreened gravel, followed by 220 mm of best seasoned gravel that can be found contiguous with the line of the road.

The techniques in Figures 3.5–3.7 are said to have been commonplace, although they made the passage of wheeled vehicles very difficult. For example, grading material to the centre of a road would force traffic to use roadside ditches.

There were also *wavy roads* with transverse hillocks in the pavement intend to ensure that some of the pavement was relatively dry. Despite passionate support in the mid-18th century, practical experience meant that they were obsolete by 1780.[100] A wonderful further account of the badness of English roads prior to the 19th century is given in the first six chapters of Smiles' Lives of the Engineers.[101]

Swamps and bogs are at the far end of the less-than-firm ground spectrum and ways of crossing them are discussed in Chapter 5. In an ideal world, the prime solution to water-logged roads would be to raise the pavement above level of the surrounding ground and to provide longitudinal drains beside the pavement. However, this requires considerable effort and organisation and experience shows that these were rarely applied to ordinary roads. Instead, as we have just seen, a range of much cheaper and more expedient measures were used, all with a notable lack success. Furthermore, although raising a pavement above the level of the surrounding land to improve drainage was usually possible in the countryside, it was often impractical in towns where buildings abutted the roadside, particularly as it was common practice for the various wastes from roadside properties to be poured onto the adjoining road.

One solution to the impassable road problem was to somehow restore the road pavement to a trafficable condition. This key issue of pavement management will be explored in Chapter 7. However, readers might find it useful to first take the Detour in Chapter 4 to learn the basis for some of the technologies that came to be employed in pavements over the millennia.

Chapter 4

The essential paving properties of soil, sand and stones

This technical detour is for all those readers who are not well versed in the intricate and somewhat arcane technologies of soil, sand and stone. Of course, such readers can happily ignore the chapter if they are prepared to accept the authors' many later categoric assertions on such matters.

Soils play a prominent role in the life of most pavements so it might be useful to share a basic understanding of their relevant properties. The first guide to the usefulness of a material is any record of prior pavement experience with the material. Probably the next most relevant measure of the usefulness of a soil or stone in a pavement is the size of the soil or stone particles. Internationally, there is now a *unified soil classification* system[102] which classifies soil components as stones below about 60 mm in size, sand, silt, clay or organic matter.

Two explicit summaries of the effect of soil types on pavement behaviour are given in Figures 4.1 and 4.2. Graphs such as Figure 4.1 were first assembled by the American Bureau of Public Roads in the 1920s.[103] They were probably useful as training tools but were too broad-brush to have much direct use in pavement design and construction. However, a very similar scheme is still used today by the American Department of Agriculture.

Fine-grained soils such as clay and silt can have a life of their own due to the water-related effects of surface tension and the electrochemical behaviour of free ions. Soils whose properties vary greatly with moisture content can be a problem for pavement builders. Pre-1950s design charts were relatively prescriptive in this respect.[104]

Clay is a very common soil. It is formed by the natural weathering of some rocks, particularly basalts, and the resulting hydrated aluminosilicates are very complex materials, physically and chemically.[106] Clay particles are microscopically small (less than 5 μm) and plate-like and carry ionic electrical charges which are negative over a plate's surface and positive along its edges. Thus, when dry, the particles stick strongly together edge-to-plate. This has immediate practical value as a dry clay mixed with other fine soil particles can ionically bind the components of the pavement layer together.

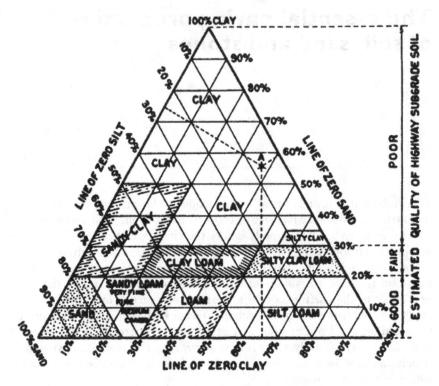

Figure 4.1 American Bureau of Public Roads "trilinear" or "triangular" soil classification
scheme developed in the 1920s.

   More broadly, these surface chemistry effects at the particle surfaces can
be much greater than any gravitational or load-induced effects. In the presence of water, the positive (H^+) and negative (OH^-) water ions attach preferentially to the plates and markedly weaken their surface-to-edge contacts.
When such clays absorb moisture, they become soft and smooth to touch
and can be pushed around, moulded and rolled into threads under a steady
force. Technically, this behaviour is described as *plastic* which more generally means that they can be pushed into a new shape without any change in
volume or increase in the pushing force. They only expand in volume, making them unsuitable for use in pavements, if their moisture content changes.
Indeed, as a clay absorbs water, its volume can change dramatically to
many times its original volume. The more expansive clays (e.g. montmoril-
lonite and bentonite) can cause the ground surface to rise by up to 60 mm
when the clay becomes wet, causing major bumps in the pavement and,
when it becomes dry again, creating cracks and other surface fissures up
to 20 mm wide. These expansive clays are therefore used with great cau-
tion in pavement making.[107] Tests to determine how much a clay sample
expands when wet are relatively simple and were developed in the 1930s.

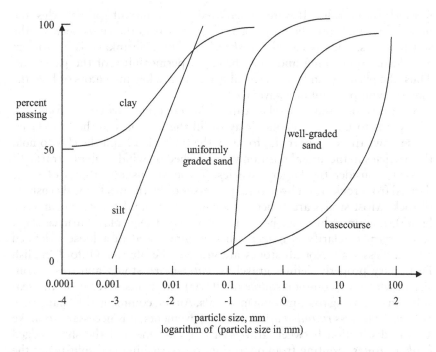

Figure 4.2 Typical particle size distribution curves.[105] "Percentage passing" refers to the percentage of particles in a sample that passed through a sieve for that particle size.

Procedures to detect both swelling and shrinking propensities in clays and fine-grained soils are now available.[108] In some cases, the expansion problem can be reduced by chemical treatment of the clays to lock up any free ions.[109] Methods to do this are discussed in Chapter 18.

On the other hand, silts, with particle sizes between 5 μm and 2 mm, have no inter-particle ionic bonding and so are dusty when dry. When wet, they gain their minimal cohesion from the surface tension of the water.

Soils are mixtures of fine stone particles, clays, silts and organic matter. As discussed above, they change their physical state from dry, to semi-solid, to plastic, to effectively liquid as their moisture content increases. In 1911, a Swedish physicist, Albert Atterberg, devised a test for determining the moisture contents at which a particular soil became plastic (its *plastic limit*) and then effectively liquid with any fine material in suspension in the water.[110] This last moisture content is called the *liquid limit* of the soil. A high liquid limit suggests that the soil mix may contain components which will not perform well in service. The *plasticity index* is the difference between these two limits and is a measure of the cohesiveness of the soil particles. Plastic behaviour is a double problem as not only will it led to a rut forming due to the passage of wheel loads but also the plastically displaced material will have to find another "home" usually by pushing the adjacent pavement upwards into a ridge. If water is removed from a mix of soil particles, the mix will shrink until the only remaining water is in the voids between the soil particles. This is called the *shrinkage limit*. Shrinkage is a problem as it leads to cracking and thus increased permeability of the pavement. Thus, the plasticity index and shrinkage limit are key indicators of how the material will perform in a pavement.

Mixtures of sand and clay can make effective low-cost pavements. The key is to have just enough clay to fill the air voids in the sand layer. This allows the sand particles to carry the traffic load and the clay to hold the particles in the lateral position when pushed by wind, water or traffic.[111]

Next, consider the larger particles. These are usually pieces of stone derived from rock strata by natural processes or by the mechanical crushing of rock. Most stones are either sedimentary (e.g. sandstone), igneous (e.g. basalt), metamorphic (e.g. hornfels) or artificial (e.g. blast furnace slag). A key term to clarify is *gravel* which is often used in a loosely defined and careless way. Not all stones are gravel. The Shorter Oxford English Dictionary properly defines gravel as *water-worn* stones and this is consistent with their common source in alluvial sediments, river-beds and sea-sides. In reviewing road-making materials, Aitken commented[112] that *gravel is all more or less rounded and worn*. In pavements, it is necessary to make a clear distinction between gravels at one extreme and the sharp-edged *broken stones* resulting from quarrying, rock-crushing and splitting at the other extreme. For road-making, gravels are very poor as they are easily squeezed out of position, and broken stones are very good as they are easily compacted into stable layers. Confusingly, the term "sharp gravel" has been used for normal rounded gravel that has been washed clean of dirt.[113] The Shorter Oxford English Dictionary does not include this use, and Phillips may have misinterpreted what he had been told. In England, stones suitable for pavements are called *roadstone*. Flaky, elongated pieces of stone also perform poorly in pavements as they also are difficult to compact into a stable layer.

Random mixes of gravel and clays were often used as, although ineffective, they were relatively easy to put in place. The lessons were slowly learnt, and by the end of the 18th century, some English road-makers were specifying "sifted and cleaned gravel".[114]

A side benefit of stone-breaking was that it detected stones which would quickly degrade under traffic. The first paving use of compacted broken stone may have arisen from observations in quarries where masons producing large stone blocks also created many stone chippings.[115] Pieces of broken stone are readily compacted into a useful surface by the wheels of wagons hauling the larger pieces of stone. We will see in Chapter 13 that a similar sequence occurred with the European "discovery" of asphalt paving early in the 19th century. A hint of this quarry-based origin of the use of broken stones in road-making is given in a 1562 Act of the English

Parliament which gave[116] parish road "supervisors" the right *to take away the ... smallest broken stones of any quarry ... lying within the parish ... for the amendment of their ways*. Technically, the term *angularity* is often used to describe the shape of broken stones.

Masons also produce cubical stone blocks for segmental paving (Chapter 10). These are commonly called *dressed* stones.

Local stones are often used for making pavements for low-cost roads, where transport costs would make it uneconomic to move large quantities of stone very far from its source. Such stones must be carefully assessed for their inter-related properties of strength, hardness and durability as, although they might appear inherently strong and tough, this will not necessarily be the case in service. For example, stones may break into flakes when crushed, be brittle and shatter under impact, be easily abraded, polish under traffic, disintegrate when wet or decompose chemically when exposed to air or water. Although these behaviours can sometimes be detected in advance using prior experience, others can be quite deceptive. A hard stone might still polish under traffic and therefore produce a slippery surface. Some basalts containing "secondary minerals" will be very strong *in situ* but will decompose to clay and small particles if exposed to the weather. The secondary minerals – often green in colour – usually can only be seen under a microscope. Adhesion is also important as bitumen will not adhere well to stones with a high silica content (so-called acidic rocks). Over time a range of tests has been developed to predict the behaviour of stones within a pavement.[117]

In selecting a stone type for use in a pavement, its durability is first addressed by using the source rock which appears in its in situ location to be durable and free from deleterious materials or defects. It is then defined by the manner in which it fractures during quarrying and by whether its surface is easily scratched. Longer term behaviour can be more difficult to detect in advance.

Impact resistance is assessed by measuring how much fine materials are produced when the stone is hit by a hammer. Similarly, fine material produced when a stone is crushed in a compression testing machine is used as a measure of its crushing resistance.[118] Abrasion (or *attrition*) in service was particularly important in the 19th century when most traffic consisted of iron-shod hooves and iron-rimmed wheels. Paving surfaces using soft stones could be quickly eroded. Even seemingly hard rocks were found to polish after just a year or so of heavy urban traffic.[119] Initially there were some attempts to classify the abrasion resistance of the stones from each quarry. Then, there were efforts to determine properties by microscopic examination of pieces of stones. The variability of the stones meant that these approaches were rarely able to predict properties in advance. They were replaced by various simple grinding tests which were used to assess abrasion resistance of stones being considered for pavement use. In 1869, the French authorities introduced the Deval test for measuring the abrasion resistance of broken stones. A sample of the stones was rotated inside an eccentric cast-iron cylinder and the amount of powder produced after a given time was compared with the quantity produced by stones which had previously been found to be acceptable in service.[120] The micro-Deval test developed in France in the 1960s is a modern version of this test.[121]

Deval-type tests can be very slow, and other tests were developed in England and America. The most widely used current test is probably the Los Angeles Abrasion test adopted in America in 1937, despite it being very noisy and sometimes giving a poor representation of how the stones would perform under traffic. The Los Angeles test follows the Deval test in that the samples are still placed in a large rotating drum. However, 12 steel spheres are also added and the acceptability criterion is the change in the stone size after 1,000 revolutions at about 0.5 Hz.[122]

When stones are used in a pavement layer without any glue (such as bitumen or cement) holding the stones together, the pavement is said to be *unbound*. In unbound pavements, the individual components act independently, although they may squash into each other and transfer loads downwards when the pavement is compressed by traffic. A simple measure of such a pavement is the density of the pavement layer. This will be less than the density of the individual pavement particles as there will be voids filled with air and water between the particles. Most unbound pavements gain stiffness from the lateral confining pressure of the adjacent pavement as materials tend to expand laterally when compressed (known as the Poisson effect).

The voids in an unbound pavement are reduced by compaction during construction and by subsequent traffic loads. This relationship between compaction, moisture content and density is a key predictor of the performance of unbound pavement layers but did not come into prominence until experience with the compaction of the walls of earthen dams in the 20th century. Specifically, in 1929 Ralph Proctor of the Los Angeles Bureau of Waterworks showed that there is a maximum density that can be achieved with a given moisture content and amount of compaction, and the test to determine these values is now known as the Proctor test. In particular, the *optimum moisture content* is important as at this moisture content, the soil attains its maximum density for a given amount of compactive effort. It will always be less than the moisture content at which the soil is saturated.

Compaction can be quantified by laboratory testing to measure the achieved density. This once required taking core samples which was slow, affected the work process and possibly compromised the impermeability of the pavement. Today there is a range of devices, such as nuclear density meters, that permit real-time non-destructive testing of achieved densities.[123] As traffic loads increased and as construction equipment improved, it became necessary to increase the "given amount of compactive effort".

Moisture contents of paving and subgrade materials are major factors in determining their performance. Materials placed at the optimum will usually stay close to that level due to the close-packing of the material, allowing little opportunity for water to enter or leave. Indeed, it is commonly found that there is an *equilibrium moisture content* which will be sustained under a sound pavement over its life.[124] Thus, if a pavement is built wet, it will probably stay wet until its premature failure.

Assuming that all the materials being used are the same, variations between mixes depend on the grading of the stone being used. If the stones being used are all of the one size, the stone batch is generally called *gap-graded*. A mix is called *open-graded* if the gap in sizes is between the largest and smallest stones. Both open and gap grading create voids in the stone mix. When the stone layer is compacted, there will still be voids (interstices) between the stones, even though they will be in contact and able to transfer loads from stone to stone. This situation is accentuated when broken stones with sharp edges are employed. A *well-graded* mix has a range of stone sizes so that the void space and the subsequent filler or mortar volume are all minimised. Thus, a pavement using well-graded broken stones would be easier to roll and compact, have a smoother surface, be less prone to frost heave (Chapter 1), be much more resistant to water penetration and be stiffer under load than the alternative mixes.

The stone proportions and sizes needed to provide well-graded (and therefore dense) mixes are based on the work done by Arthur Talbot in 1923 to aid the design of concrete mixes.[125] His analysis examined the theoretical packing of spheres and produced the now widely used grading equation:

(% passing size d)/(% passing size D) = $(d/D)^n$ where n is between 0.3 and 0.5

As macadam pavements came into widespread use (Chapter 9), specific requirements developed for the stones to be used. Particle size in McAdam's time (Chapter 9) was typically in the range of 40–60 mm. Thus, the requirement was effectively all stones smaller than 60 mm and none smaller than 40 mm. This changed when mechanical crushing was introduced (Chapter 9) as this machinery could control maximum size but not minimum size as two or more stone fragments could be crushed together, thus producing smaller particles. This consequence required the broken mix to be passed over one or more screens (or sieves) to retain only the particle size required. In the simplest case, one screen would have an opening of 60 mm so that over-sized particles could be discarded, or reprocessed, followed by a 40 mm screen so that the material required would be retained and particles which were too small could be discarded.

As the importance of a well-graded mix came to be realised, stones were routinely sieved to ensure that the required grading of sizes would occur. A typical *grading curve* was shown in Figure 4.2 which extends into the finer materials, sands and clays found in subgrades. The behaviour of these smaller particles was discussed earlier.

The engineering use of stress and strain should also be explained. *Stress* is load divided by the loaded area and will usually imply that the load is applied normal to the surface. *Shear stresses* arise when the load is applied parallel to (along) the surface. *Strain* is the deformation caused by the load divided by the unloaded length. If all this strain disappears when the load is removed, the material is elastic. Stress and strain are simple concepts on a large scale – e.g. tens of millimetres – but begin to lose meaning and become difficult to measure at the smaller scales associated with the internal structures of typically discontinuous paving materials.

The *modulus of elasticity* is the applied stress divided by resulting strain. It is a direct measure of the *stiffness* of a material. If the modulus of elasticity is constant, the material is considered *elastic*. For an elastic material, it is relatively easy to calculate the strains caused by a given stress. However, the elastic modulus of asphalt drops as temperature increases. Pavement engineers sometimes use the term *resilient modulus* which can be either the modulus of elasticity associated with the rate at which the stress is applied or, in a confusion of concepts, the ratio of the applied stress to the strain recovered when that stress is removed from a non-elastic material. Various algebraic approximations to the stress–strain behaviour of pavement materials have been suggested.[126]

Most paving materials – including asphalt – are not perfectly elastic, and some permanent deformation will occur with each significant wheel pass. A further concept is *creep*, which describes the strains that occur in some nominally elastic materials such as asphalt when subjected to a constant stress over a long period of time.

Eventually, as the magnitude of the stress or the number of stress cycles increases, the material will fail catastrophically by cracking if the stresses are tensile or squashing if the stresses are compressive. The associated *strength* of a material is a much more nebulous concept as most materials will fail gradually rather than catastrophically and so subjective definitions are needed that relate to a particular unacceptable circumstance. For example, when does a rut represent pavement failure? The rut is the product of permanent deformation of the pavement usually as a consequence of an increment of compaction caused by each passing traffic load. The damage occurs incrementally and slowly unless the rut creates new problems such as permitting water to enter the pavement. At the other end of the damage spectrum, the tensile stresses that cause cracking will also grow incrementally under traffic until they trigger some other undesirable effect. This type of damage is called *fatigue* failure. While some cracks are due to surface shrinkage, structural cracking usually begins at the base of a pavement layer or in the subgrade where the tensile stresses are highest. When such cracks grow over time and reach the surface, they are called *reflection* cracks.

Within the pavement, a stone must be strong enough to resist crushing or breakup under impact forces during construction and subsequently when under traffic, particularly at contact points between individual stones. At the pavement surface, abrasion resistance is also required. The size, shape, strength, stiffness and hardness of stones are relatively easy to assess, particularly during the stone-breaking process. Conditions were generally more demanding before the era of asphalt and concrete and when the traffic was predominantly horse-drawn iron-wheeled vehicles. Durability over years due to weather and chemical attack and abrasion under traffic are much harder to measure, but useful tests were gradually developed.[127] This issue is pursued further in Chapter 19. Some stones when mixed with cement to make concrete have been found to slowly react with the cement, and the product of the reaction has caused serious long-term damage to concrete. The problem is known as *alkali-aggregate reaction*.

As will be discussed in Chapter 10, in the late 19th century there had been some compression testing of stone paving blocks.[128] A 20th-century innovation at this end of the test spectrum was the *repeated load triaxial* test which attempts to replicate the service behaviour of a pavement by subjecting a laterally confined cylindrical sample of the pavement to repeated loads on the two ends of the cylinder.[129] The loads in size and frequency simulate the expected traffic loading. Confining the specimen replicates the horizontal stresses produced by the adjacent pavement material as it tries to expand laterally (the Poisson effect). The lengthwise deformation of the cylinder indicates the propensity of the pavement to rut under service conditions. However, by its very non-uniform and non-elastic nature, it is difficult to apply the classical methods of stress-and-strain analysis to an in situ pavement layer (Chapters 16–19). Some so-called triaxial tests had no horizontal restraining stresses and these are better called unconfined compression strength tests.

The relative rutting-related damage done to asphalt pavements by wheels carrying different vertical loads is estimated by an empirical formula known as the fourth power law which says that a wheel load which is m times greater than some given load will do m^4 times as much damage as the given load. For concrete pavements, the power is greater than four. The overloaded wheel will thus reduce the time for a pavement to reach a terminal condition by a factor of $1/m^4$. For example, if the overload was twice as big as the design load, the life of the pavement would be $1/2^4 = 1/16$th of its design life – a dramatic effect. For fatigue cracking, the power may be closer to 2 rather than 4.[130] For weaker subgrades, the power might be as high as 12.[131] Fatigue in most of the materials follows similar power laws. A specific examination of the damaging effects of overloading led to the strong recommendation that it is better for pavement managers to concentrate on strategies to prevent overloading, rather than to detect and then penalise or tax overloaded vehicles.[132]

End of detour.

Chapter 5

Paving ways through swamps and bogs

Although many of the pavements built on firm ground and described in the previous chapter were less than adequate, an even more challenging situation was encountered when the ground was a swamp or a bog and far from firm. As swamps and bogs were frequently associated with fertile areas, there were often social and economic imperatives to find a way through the challenging ground.

In forested areas, a well-used solution was to place logs on top of the swamp and allow them to sink under their own weight. Logs would continue to be placed until their upper surface was above the ground level, due either to the logs resting on the bottom of the swamp or the whole "structure" floating in the swamp. In the latter case, it was frequent experience that structure would sink further when trafficked and would require a fresh new layer of logs to be placed at the surface. Pavements of this general nature across wet or boggy ground are commonly called *causeways*. In 1474, Maud Heath used her life savings to build a causeway, partly through the bog between her house and the Chippenham market where she sold her eggs. It is still in service and still carries her name.[133]

In the simplest application of the method, the surface layer of logs was placed transversely, and the appearance of the surface (Figure 5.1) led to such roads being called corduroy due to their resemblance to the textile of the same name. Peat bogs throughout Europe, for example near Glastonbury in Somerset, contain long-sunken but well-preserved current examples of such roads. Various timber causeways dating from pre-Roman times have been located at many locations throughout Europe.[134] Many applications in these situations have used brush or woven panels of branches instead of or in addition to logs.[135] Proctor's 1607 solution involved placing brush or branches within horizontal wooden frames which spanned between longitudinal drains on both sides of the road. A more brutal solution occurred in 1665, when the Bishop of Münster during a war between Germany and Holland built a 6 km road though Bortanger Moor by supplementing brushes with pieces of houses destroyed in nearby villages.[136] More peacefully, it was also found that the pliable living branches of some trees, such as willows, would form a strong and coherent living root mass, when bent

Figure 5.1 20th-century corduroy pavement near Apollo Bay, Victoria, Australia.[137]

down and then placed horizontally within the pavement. The branches formed a coherent living basket weave across the road which was then covered with suitable soil. Logs were also used longitudinally to fill deep ruts created by wheeled traffic.

The 690 km long Perspective Road (Perspektivnaia Doroga), built between Moscow and St Petersburg between 1722 and 1746 relied on long stretches of corduroy pavement.[138] Some 28 layers of logs have been found under one portion of pavement (Figure 5.2).[139] It is estimated that one layer of logs was placed about every 20 years.

Various earlier (pre-1500 BC) versions of the method have been found near Prague, around "lake" villages in Switzerland,[140] in the Pamgola swamps in Hungary, between Magdeberg and Scherwerin in central Germany and elsewhere in the flat and swampy loess and peat lands of northern Europe,[141] southern Holland and Glastonbury.

Indeed, excavations in peat-filled basins near Glastonbury in the Somerset Levels have provided many insights into the construction of pavements in swamps in ancient times.[143] Some of the Somerset roads were stabilised by vertical logs located at about a metre spacing along both "edges" of the corduroy road.[144] A later pavement development in the same area was revealed during excavations of the Sweet Track where the construction sequence in about 3800 BC was to first put a single row of logs on top of the marsh on the line of the intended path. About every metre, pairs of pegs were driven obliquely into the marsh above the line of horizontal logs. Timbers for walking on were then supported by the V's created by each pair of pegs.[145]

Figure 5.2 Corduroy road near Novograd exposed during an archaeological excavation.[142]

It has been estimated that a kilometre of path would have required 20 km of timber and about 50,000 m long pegs. Thus, the construction effort was substantial.[146] A related technique was to produce the horizontal walking surface by weaving thin branches into a mat known as a *hurdle*. This method is illustrated in Figure 5.3. Such techniques used in the Somerset bogs have been investigated and documented by the Coles.[147] Subsequently, the Romans used log roads extensively in swampy regions, calling them "long bridges" (*pontes longi*) and to this day, the method has continued to be applied in remote forested regions.[148]

A more structural solution would have been to hammer logs vertically into the swamp until they could be driven no further. Such logs are called "piles", and a path could be made by using horizontal timbers to bridge between the piles. A Roman causeway built across a marsh in the Medway valley in Kent used vertical oak piles over a metre long with pieces of wood laid horizontally between the piles.[150] A structural extension of this approach was found by Mertens in marshy ground on the Roman Via Mansuerisca in Belgium.[151] Another Roman method is discussed in Chapter 6.

Figure 5.3 The "Walton Heath track" – using carefully linked branches to create a "hurdle" pavement in a swamp in Somerset, c2300 BC.[149]

A well-preserved bog road was found in the Bog of Allen in central Ireland. Dated at before 50 BC, it was described as follows:[152]

> Beams of oak were laid down across the line of the road; then others were laid above these, parallel with the line of the road, and so in turn until five layers of beams had been deposited. A bedding of gravel and sand was laid on the topmost layer, and above that there was a remarkable pavement of birch rods, each about an inch in diameter......It is probable that sods were laid on the birch rods to form the pavement.

In another variant of the corduroy road, longitudinal planks were placed on top of the transverse logs to provide a good running surface for wheeled vehicles (Figure 5.4). The method was well-used in Russia and was introduced into Canada in the 1830s.[153] In 1835, Darcy Boulton developed Yonge St in Toronto using a version of the method that came to be called a *farmers' railroad* and was subsequently widely marketed in North America and Australia. The first plank road in the United States was a 30 km toll road built in 1846 between Syracuse and Oneida Lake in New York State.[154] Typically, the method cut logs in half longitudinally. They were then placed in the future wheelpaths with their flat surface uppermost, horizontal and slightly below the ground level. Flat planks were then placed side-by-side transversely across the logs and pounded down until they bore on them. Such roads had a life of about 10 years as the timbers decayed and rotted.

Figure 5.4 Plank road between El Centro, California, and Yuma, Arizona.[156]

Well over a thousand kilometres were constructed in North America but, by 1903, only a few plank roads remained in service.[155] Those that remained were mostly in desert areas where rotting was minimal.

The use of bundles (called bavins, faggots or fascines) of small branches, brush, heather, ling or gorse[157] to improve soft subgrades was commonplace. There are reports that King Edward III in the 14th-century England could only reach Parliament after bavins were used to fill holes in the approach roads to Parliament.[158] The use of bavins was widespread in the 18th-century England.[159] The Kensington, Fulham and Chelsea Turnpike Trust purchased 36,000 bavins in 1726 and 37,000 in 1727.[160] The bavins were often then covered by larger branches following the hurdle method shown in Figure 5.3. In the latter half of that century (1750–1790), a well-known Yorkshire road-builder, John Metcalf,[161] successfully extended the technique on 15 km of road between Huddersfield and Manchester by using rafts of bavins and hurdles to literally float a pavement over a troublesome swamp. The material placed on top of bavins was often obtained during digging of longitudinal drains along the roadside. A well-used 1850 English textbook on road-making recommended[162] that *where the ground is very soft, a layer of (bavins) should be laid over the surface of the ground before laying on the road materials.* Bavin-based methods have continued to be used to the present day.[163] Bavins have been found to last long if permanently wet but to soon rot if they are allowed to be alternatively wet or dry.

In the 1840s, it was reasonably, but somewhat impractically, argued that the best way of crossing boggy ground was to provide good drainage of the boggy subgrade and then to cover the subgrade with about 400 mm of "stiff clay" before placing the pavement on top of the clay layer.[164]

Stone paving was used in roads crossing somewhat more resilient swamp land. Excavations of a road across the Hulebæk River between Præstø and Tappernøje in Denmark showed a road, possibly of Roman origin, paved with flat stones. In medieval times, another pavement of large and small stones had been built on top of it and it now functions as a submerged ford.[165]

Chapter 6

Roman pavements – a major advance in pavement quality and extent

In the four or five millennia over which pavements have been used by many societies in many lands, Roman roads and their pavements represent exceptional peaks in both their extent and their quality; peaks that have only been reattained in the last 200 or 300 years. This remarkable Roman achievement therefore deserves careful attention.

The Roman Republic was established in 509 BC and, by the 3rd century BC, Rome controlled the entire Italian peninsula. The streets of Vetulonia about 200 km north of Rome were stone-paved by about 400 BC. The Romans were probably influenced by Carthaginian pavements and by Minoan construction processes (Chapter 2), copying their use of cement which they enhanced by adding volcanic tuff which, if finely ground, also reacts with water to form a mortar. The tuffs came from the Pozzuoli region near Naples and such materials today are called *pozzolans*. This procedure appears to have been inherited from the Greeks and Etruscans. When the Roman Empire disappeared in about 500 AD, the manufacture and use of cement disappeared with it and was not rediscovered for another millennium. Cement is discussed further in Chapter 11.

The Republic began building arterial roads in 334 BC with the construction of the Via Latina from the Roman gate now known as Porta San Sebastiano to Capua in the south-east, a distance of about 200 km. Before the Republic expanded, Rome followed Greek practice and required young men to devote up to 2 years working on roads and other municipal projects. The road work nevertheless exhausted the public treasury.

By 312 BC, the Via Latina had been replaced by the much straighter Via Appia (Appian Way) named after the censor Appius Claudius who had initiated its construction and then took great glory in the outcome. The title Via was reserved for major roads. Subsequently, the Via Appia was sometimes called the Regina Viarum (Queen of Roads). By 264 BC, the road had been progressively extended a further 400 km across the Italian peninsula to the Adriatic port of Brindisi. The Appian Way had begun as an earthen track, and the first Roman improvement was to apply a surface of small stones. It was paved with stone slabs between 295 BC and 123 BC. The Way was partly reconstructed in 130 AD and was in routine use until the 6th century and remains in restricted use to the current day. Given its life of over two millennia, it shows surprisingly little wear. There are many excellent photographs of the Appian Way available online at sites such as Wikipedia Commons.

The incentives for building the Appian Way were mirrored in all the other subsequent Roman roads. First, they allowed the Roman army to capture, control and then protect the Empire's growing territories. Second, they allowed Rome's very centralised government to administer and tax all of its distant empire. Third, they allowed the local administrators to send to Rome large numbers of newly acquired treasures and slaves and to provide the imperial treasury with reliable and regular taxation revenue. Thus, the prime use of the Roman roads was to facilitate the rapid and effective deployment of the army, followed by the unimpeded return of wagon-loads of booty and the effective supervision of remote administrators. The process was aided by the fact that, during times of relative peace, road-building also provided gainful employment for otherwise troublesome soldiers and their many prisoners of war. Indeed, in his detailed account of Roman roads Nicolas Bergier said that a key Roman "raison d'être" for their pavements was *to keep the multitudes out of idleness*.[166] For all these reasons, Roman roads had clear objectives and served profitable purposes, factors which made devoting resources to their construction and maintenance a readily accepted burden.

Meeting these objectives required wide roads, safe roadsides, all-weather pavements, pavements smooth enough for wheeled vehicles, and gradients flat enough to enable the passage of heavy hauled wagons. This was an order or two more challenging than that posed by the street pavements discussed in Chapters 2 and 3. Roman roads are still being located – it is currently estimated that their total length was well in excess of 100,000 km.

The Chinese Empire at about the same time faced and addressed similar challenges.[167] The roads they constructed in the alluvial planes were commonly of earth compacted with hand-held rammers and, where needed, covered with flat stones.[168] In Chinese road-building, *a labour force of twenty to thirty thousand men were put to work building a road*.[169] The Japanese road system began to develop in the 7th century AD with about 6.5 Mm of arterial routes brought into operation over the next century and reflecting the central power of Kyoto, the capital at the time.[170] The Roman achievements are currently far better documented than those of China and Japan and so will be the major subject of this review.

The major Roman roads would at least permit two carriages to pass which implies a width of at least 7 m although many were of lesser width. At this point, it must be noted that the Romans themselves left no written records of their road-building practice and a wide variety of cross-sections have been located, as illustrated by the measurements of two British Roman roads in Figure 6.1.[171]

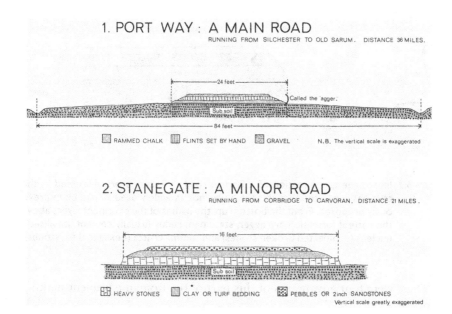

1. PORT WAY : A MAIN ROAD

RUNNING FROM SILCHESTER TO OLD SARUM. DISTANCE 36 MILES.

24 feet

Called the 'agger'.

Sub soil

84 feet

RAMMED CHALK   FLINTS SET BY HAND   GRAVEL   N.B. The vertical scale is exaggerated

2. STANEGATE : A MINOR ROAD

RUNNING FROM CORBRIDGE TO CARVORAN. DISTANCE 21 MILES.

16 feet

Sub soil

HEAVY STONES   CLAY OR TURF BEDDING   PEBBLES OR 2inch SANDSTONES

Vertical scale greatly exaggerated

Figure 6.1 Bodey's cross-sections of two British Roman roads.[172]

The Romans were able to build such roads because the need was great, the necessary human and physical resources were available and they had access to many skilled stonemasons. In addition, they had time and space to refine their techniques by trial and error and the steady application of common-sense unhindered by local preferences and prejudices. Although most Roman roads eroded over time, due to traffic, weather and locals plundering the stones for their own private purposes or to prevent outsiders for accessing their land, a few examples of a complete pavements still exist. One such example is the Fosse Way south of Radstock in Somerset which was "opened" archaeologically in 1881.[173]

An illustrative cross-section of a Roman road is shown in Figure 6.2. The process the Romans used was to first construct longitudinal drains along the future edges of the road. They would then excavate the proposed road alignment down to firm ground. Next, this would be bought to the required subgrade profile by using stable, local stony material, often provided by the excavation of the longitudinal drains beside the road. It was compacted into a firm layer by using hand-held timber rams (Figure 6.4), and the surface then levelled with sand. This embankment-style layer placed on the firm ground to provide the subgrade for the pavement was called the *agger* and could raise the pavement by up to a metre above the level of the surrounding land.

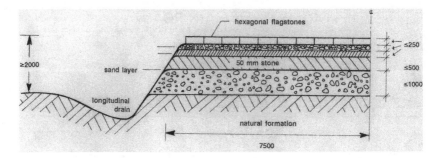

Figure 6.2 Illustrative cross-section of a Roman road.[174] Recent work discussed in the text has suggested that this form was not as widely used as had been previously proposed. From the bottom up, the names of the pavement layers above the natural formation are agger, statumen, rudus (usually cement stabilised), nucleus (compacted broken stones) and summa crusta (hexagonal flagstones).

There were four advantages of this part of the Roman pavement-making process:

- the deep longitudinal drains kept water away from the pavement and drained water from the pavement (the various damaging effects of water were discussed in Chapter 1),
- making the drains often provided material for the agger,
- the agger raised the level of the pavement and this would help to avoid the road being flooded, and
- raising the pavement made it easier for road users to defend themselves from roadside attacks.

In swampy or marshy areas, timber was used to produce a "firm ground" *agger* equivalent. This was done by building a horizontal timber frame, *contignata*, which was covered with brush or reeds and then effectively floated into position in the marsh. Its stability was sometimes aided by also using timber as vertical piles. Other versions of the process are discussed in Chapter 5.

The pavement layer placed on top of the subgrade was called the *statumen* and was typically about 500 mm thick. It was prepared by stonemasons placing stones at least 50 mm thick in tightly jointed horizontal layers. Cement mortars were sometimes used in the joints. Stone was the common building material of the era, and so skilled stonemasons were relatively easy to obtain. The statumen was the major structural layer and also prevented the vertical movement of small particles within the loaded pavement. An English commentator in 1776 described the Roman Fosse Way south of Ilchester in Somerset as *composed of flat quarry stones of the country, of good breadth, laid edgewise, so close that they looked like the side of a wall broken down.*[175] Statumen layers have been the most visible remnant in many surviving Roman roads.

The layer immediately above the statumen was the *rudus*. This was a layer up to 250 mm thick comprised of smaller stones bound together by a cement mortar. It filled interstices in the surface of the underlying statumen and produced a smooth impervious upper surface.

The next layer was the *nucleus* which could also be up to 250 mm thick. It was also mortared but used smaller stones and was more carefully compacted. Compaction was usually done with timber stumps about 1.5 m long. When iron became more plentiful, [176] iron straps would be used to bind the end of the stump. The stumps were rammed down by workers following the beat of the overseer's drum. The Romans sometimes achieved the desired compaction levels by then using large cylindrical stones working as rollers and moved by slaves or oxen pulling on timbers or ropes attached to the rollers' shafts.

If the traffic was heavy, or the locale important, a surface layer called the *summa crusta* or *pavimentum* was placed in a manner similar to the earlier Mesopotamian pavements discussed in Chapter 2. The carefully fitted hexagonal flagstones were approximately 600 mm by 200 mm thick. These can be seen in photographs of the Via Appia, available on websites such as Wikipedia Commons. Their use reflected the common availability of skilled stonemasons. Examples of where marble was used for the flagstones have been found in some urban areas.[177] At the other extreme, Davies has observed that such paved surfaces were rarely used on English Roman roads.[178]

The pavement surface sloped laterally to ensure that water drained quickly from the surface and into the longitudinal drains. The finished surface was typically about 15 m wide with the paved central 5 m strip devoted to wheeled traffic. Kerb stones were often also employed. The two outer 5 m strips were sometimes composed of broken stone which was easier for foot and hoof traffic. At about the same time, the Chinese were using a similar three-part "longitudinal strip" system and much later Napoléon also employed the method, although one of his outer strips was paved with earth to aid the cavalry horses. Napoléon was also particularly interested in the paving of the floors of the stables that housed his Army's horses.

Roman roads could be massive structures by modern standards and their engineering skill lay in their construction rather than their design. They were built to last, although, as will be discussed below, many were deliberately destroyed. Nevertheless, many have lasted in one form or another to present day, often used as local ways or hidden beneath shallow layers of soil. With others, the surface has been repaved with slabs or the upper courses have long since washed or worn away (Figure 6.3) or they now display wheel ruts worn in the stone surfaces over many subsequent centuries.

Not all stone-paved roads date from Roman times. Figure 6.4 shows a possible rutted Roman road across Blackstone Edge leading out of

Figure 6.3 Roman road in Spain at Puertodellio pass in the Sierra de Gredos in the 21st century. What is now visible is probably the statumen as upper courses would have long since washed or worn away.[179]

Lancashire into Yorkshire. Addison states[180] that *it is probably the best preserved section of Roman road in Britain*. However, Davies has subsequently questioned whether it is a Roman road and suggests that the rut, at least, might date from the turnpike era.[181]

Figure 6.4 Rutted stone-paved road across Blackstone Edge leading out of Lancashire into Yorkshire[182]

The hierarchy of pavement layers shown in Figure 6.2 and discussed above was proposed by Bergier in his 17th-century review of Roman roads and has been widely repeated.[183] However, Chevallier in the mid-20th century noted that *the manner in which the roads were built varied a great deal, even along the same route.*[184] For example, given local circumstances and the quality of the agger, parts of the statumen, rudus, nucleus and pavimentum could be omitted. No Roman road-making manual has ever been discovered, and Bergier was greatly influenced by the work of Vitruvius which dealt with the construction of the floors of buildings, not roads.

Thus, the practices described here should not be seen as based on some common Roman road-making standard, and Davies gives an interesting review of the various pavement structures that he has encountered in England.[185] He notes that on the one hand some pavements were clearly added to over time and on the other hand that some of the lower layers were not used when the natural subgrade was adequate for the purpose. In examining the Roman "Pilgrim's Road" in Turkey, archaeologist David French recorded that *There was no suggestion of a complex substructure to the road surface. Edging stones seem to have been placed directly on the existing ground surface; there is no evidence of trenching. The edge stones are up to 300 mm deep. Smaller stones (average size 200 mm) are then spread on the existing ground surface as a paving between the edge stones.*[186] Any smaller stones needed to provide a trafficable surface probably did not survive the passage of time.

Despite the many admirable examples of operating Roman pavements, Roman road-making and even many of their roads quickly disappeared due to a lack of maintenance as the Roman Empire began contracting after about 300 AD. Many of their roads were also subsequently partially or totally destroyed by locals seeking useful building stone or hindering the access of invaders, plunderers and pillagers. In the late 19th century, some French railways used stone from Roman roads as railway ballast.[187]

Chapter 7

Pavement management processes from medieval Europe

An ongoing problem with even the simplest of pavements has always been finding sufficient resources to manage and maintain them as for most of our travel history, there has been no steady funding source for this task.[188] The Roman roads discussed in Chapter 6 were quite exceptional – they had clear objectives and profitable purposes which made devoting resources to their construction and maintenance a readily accepted burden. However, conventional roads are used by travellers who have no direct obligation to fund their upkeep and pass through land whose owners benefit in only a relatively small way from the operation of the road. Over the millennia, many methods have been tried to overcome this problem and nearly all have failed until finally in the 20th century, the taxation of motor vehicle fuel provided the basis for a reliable and simple revenue stream.

The demise of the Roman Empire dramatically reduced the need for trade and travel. This led to minimal road building and maintenance, and some active destruction of Roman roads was discussed in Chapter 6. The following millennium also saw communities retreat into self-contained subsistence. Slowly societies redeveloped, trade between communities became more commonplace and the demand for usable roads increased. Timber and stones were still the main materials used for paving. The earliest signs were in France where some east-west routes were added in the 6th century and where Charlemagne in the 9th century reconstructed some Roman bridges.[189] Charlemagne also made some ineffective efforts to reinvigorate the Roman roads in France.[190]

In Anglo-Saxon times, it was widely accepted that the local rulers had three obligations – the *trinoda necessitas* – repairing roads, building fortifications and serving in the army. Travel began to gradually increase late in the Middle Ages. In 1184, Phillip II paved some Parisian streets to reduce the stench outside his palace, an issue revisited in Chapter 12. More widely, in 1222 he used his Royal power to reactivate French road construction, beginning with the resurrection of a Roman road. Construction of a radial road system based on Paris was pursued as it was associated with a growth in central power supported by a strong national army. Wagon traffic also increased steadily. In 1280, Judge Phillipe Beaumanoir defined Roman roads in French law as the fifth and greatest class of the nation's roads. Throughout Europe, roads were notoriously bad in the flat, soft and relatively stoneless lands in the north and relatively good in the stonier lands, such as Switzerland and Bavaria, to the south.[191]

In England, day-to-day administration was very decentralised and the obligation to maintain a road rested largely with the owners of the land adjoining the road. This was reinforced by law and tradition. For instance, the first major English law text, de Glanville's Tractatus was written in the 12th century and defined obstructing public ways as a crime. In 1285, Edward I wrote to the Prior of Dunstable noting that the road through Dunstable was *so broken up and deep*, that *dangerous injuries continually threatened* travellers. He commanded the roads *to be filled in and mended*, or else he would *apply a heavier hand*.[192] Dunstable was on a key northern route out of London. In 1294, Edward, faced with Welsh insurgency, ordered the Cistercian monks of Strata Florida in Wales to mend their roads.[193]

Nevertheless, the first practical solution to the impassable road problem was for travellers who found their way obstructed to seek firmer ground immediately adjacent to the impassable stretch. This could often be done with some immunity as law based on the concept of the Kings Highway gave travellers more rights than the land-owners.[194] As the fencing of private land became more commonplace, the tactic of seeking firmer adjacent ground became increasingly less practical. When the roads of the authors' home town in Melbourne were planned in the 1830s, the highway reservations were to be 3 chain wide (60 m) compared with one chain for normal roads. Part of the justification was that 60 m would allow the 2 m wide wagons to find firm ground somewhere within the reservation.

In the Middle Ages, stonemasons working as road-makers came to be known in France as *paveurs* and in England as *paviors*. Their craft was valued and they formed artisan guilds, with both apprentices and senior craftsmen. In London, today there is a Livery Company known as the Worshipful Company of Paviors. The Company, itself formed in 1479, believes that paviors as a group were formed prior to 1276. The first recorded use of the word *pavior* was in 1282. In 1280, the City required each of its Aldermen to appoint "four reputable men" to manage its pavements, and in 1302, the four were formally referred to as paviors.[195]

In France, skilled paveurs are mentioned in official documents of Charles VI in 1397 and, in 1400, he required all paving to be done by such paveurs and checked by independent inspectors. There were three classes of paveur: Companion Pavers, Master Sworn Pavers and four Masters of Paving Works appointed by the Government (Châtelet). Their work was covered by detailed regulations first issued in 1502. These included requiring that the blocks be sound and placed on good, clean, well-rammed sand. Despite the regulations, Parsons comments that the paveurs were *a lawless and undisciplined lot*. By 1547, penalties for poor paving had increased from fines, to flogging, imprisonment and hanging.[196] The paving guild continued to be powerful until about 1750. Drawings exist of pavement artisans working in Nuremburg in Germany in 1698. [197]

Road management was apparently ineffective in London as, in 1315, Edward II issued a writ on the Mayor of London requiring London pavements to be *repaired, cleansed and freed of vagrant pigs*. The Mayor then summoned the City's masons and required them to elect six paviors, experienced and responsible men, to undertake the task.[198] The response was not adequate. In 1339, a Parliamentary sitting was abandoned as the roads were so bad that most Members, even those living in London, could not reach Westminster. Pigs were a particular street problem, due both to their effluent and the damage done by their small hooves. A Parisian ordinance of 1349 not only banned pigs but also required officials to kill the trespassing pigs and fine their owners. In 1476, a Parisian ordinance required all citizens to pay a tax to fund street cleaning contactors.[199]

In 1346, Edward III authorised tolls on roads near Temple Bar in London and in Dublin. In 1353, he issued an order requiring paving with the work funded by adjoining land-owners and by all vehicles and laden beasts bringing merchandise into London. Later in the 14th century, Richard II formed the Guild of the Holy Cross in Birmingham. For over two centuries, it maintained various *foule and dangerous high wayes* to enhance the passage of the King's subjects marching to and from the marshes of Wales.[200]

It was usual to find that the basic methods for restoring impassable road pavements to a trafficable condition would only utilise minimal resources and so were consistently ineffective. In 1315, Richard de Kellawe, Bishop of Durham, applied a new solution when he issued a decree which offered to *remit forty days of the penances imposed on all our parishioners ... who shall help ... by bodily labour in the building or the maintenance of the causeway between Brotherton and Ferrybridge where a great many people pass by*. This was consistent with the concept of the *trinoda necessitas* but which now saw keeping roads in repair as a pious and meritorious work before God and which therefore could lead to the remission of sins.[201]

Direct local taxes for roads and bridges were called *pavage* and *pontage*, respectively. They were tried spasmodically in England from the 13th century but were difficult to manage and police. Local taxation was given legal status in England through paving Acts for specific towns, beginning with Chester in 1391.[202] These required citizens to maintain the pavement in front of their houses. This was consistent with a common legal view that abutting land-owners owned the half of the width of the street in front of their property.[203]

The materials to be used in paving were rarely specified. In 1443, there are reports of loads of "stones and gravel" being dumped on the road in front of Savoy Palace in London, raising the level of the road to the distress of people in adjacent properties.[204] It was a matter of ongoing concern to residents when their streets were paved.

Action had also been occurring within Paris. In 1348, the Provost of Paris codified earlier ancient (common) laws requiring property owners to pave the street outside their property, and in 1404, the Provost prosecuted the Abbot of Cluny for not paving outside his college. Later this was modified to have all the work on a street length done by single contractor, paid for by a tax on all the abutting land-owners.[205]

Pavements on roads outside of the towns presented a different set of problems. In England in the Middle Ages, there was a vague common law obligation for the local "Lord of the Manor" to maintain the roads within his manor. The Church also gradually picked up some road obligations in order to service its own particular needs. This avenue disappeared in 1530, when Henry VIII dissolved the monasteries. The core problem in England and elsewhere was that it was difficult to impose the task of maintaining a road on people using the land abutting the road, when they were not the primary road users and received little benefit from the road.[206]

A common solution in medieval times was to require locals to spend a number of unpaid days a year maintaining local highways. They did this under the supervision of another local, possibly the parish priest or the Lord of the local feudal Manor. Such methods, called *corvée* based on a French term for French military fatigue duty, had existed under various feudal systems. They were used in ancient Egypt, Greece, and pre-Roman Gaul, and later in the Russian empire[207] where locals had the option of doing the work or being beaten with a whip.[208]

Henry VIII codified the corvée obligation in the Highway Act of 1555 and its associated statutes.[209] Parishes were required to meet annually and elect two unpaid Surveyors of Highways to supervise and direct the required corvée labour. The Act and its subsequent amendments governed road maintenance in England for the next 300 years until abolished under the General Highway Act of 1835. The system also operated in Ireland between 1615 and 1763.[210] In England, the work obligation came to be called *"statute labour"*. It took a very different approach to the French, moving the responsibility for roads from the local feudal manor to the local parish. Every parishioner had to perform six consecutive days of annual labour on the local roads. Land-owners and the like also had to supply a hauled cart and two men. Neither the Surveyor nor the workers were paid or required to have had any prior knowledge of road maintenance. The Surveyor was to give notice of any defects found in roads every four months on a Sunday in the parish church after the sermon had ended. An alternative approach introduced in the Dublin City Ordinances in 1613 required *on Wednesday and Thursday weekly after dinner one man at least shall come from each house to do public works. Meanwhile, all shops are to be closed.*[211]

Wherever it was tried, corvée-type work was considered to be degrading and was avoided wherever possible. A typical procedure used by parishioners to minimise the unpaid time they spent on the task was to undertake it during a dry summer period when the roadside material was relatively dry and could be placed and then smoothed to achieve the desired convex cross-section (Figures 3.5–3.7). The road repair was then inspected and passed by the unskilled "surveyor", before new traffic and late summer rains could undo their repairs. The silliness of returning eroded material from the roadside to the road would have been obvious and anyone who had walked on a gravel pavement would have observed how much better the pavement was when the pavement used broken stones rather than dirt or local rounded pebbles. Nevertheless, to the locals the silliness clearly made a point or two.

Early in the 17th-century France, the Duke of Sully, under Henry IV, introduced a centralised national administration. A key task was an annual report on the condition of France's roads, including estimates of the cost of any required maintenance.[212] Priority for roads continued under the leadership of Richelieu and Colbert and their emphasis on France having a strong and deployable national army.[213] The need to fund this focus encouraged use of a national corvée system and, in 1738, it was extended to national roads (grands chemins). The need to manage the process led in 1747 to the establishment of the École Nationale des Ponts et Chaussées to train road engineers.[214] As a result, major French roads were kept in reasonable order, but the same could not be said for the remainder of the road system which still employed a feudal corvée method. The continued application of the corvée was a significant factor in the French Revolution late in the 18th century and in unrest in the middle of the 19th century. The corveé's practical problems were manifested, as indicated in the following review of French practice: *The main objection to employing peasants or paupers was that they did as little work as possible, and what work they did do was often done badly. Statute labour used for mending roads was both unwilling and unskilled in the art of road making. The system presented other problems such as the seasonal character of the work: the corvéables could be called upon only once or twice a year for few days.*[215]

Administrative and financial anarchy during the French Revolution late in the 18th century had a major impact on French roads. The corvée was completely abolished, and road administration was passed from the Corps des Ponts et Chaussées to local regions (départements) where the local administrators were inexperienced and unskilled. In the 1840s, a British commentator observed:[216] *In some places the (minor cross) roads are so bad that even riding is unsafe with any but horses and mules accustomed to their rugged and rutted surfaces.* In 1797, the new republic realised that its roads were now in a deplorable state, but no remedy was forthcoming. In 1806, Napoléon introduced a tax on salt to pay for road works, but the funds were diverted to other uses. Things began to improve after the

1830 Revolution, and a law in 1836 (*prestation en nature*) provided reliable revenue for roads based on a tax on all able-bodied males.[217]

The idea that unpaid and unskilled foremen and labourers might be able to produce effective pavements in the manner (to be described below) of the French engineers or of Telford or McAdam in England was beyond any local perception. As a consequence of the above factors, many roads over much of Europe's travel history between the Romans and the 19th century could be described as quagmires[218]. In England, the change to using skilled people only occurred in the 18th century when Telford worked directly for the British national government and McAdam for turnpike operators whose income depended on their roads being open for toll-paying users.

A list prepared by McAdam of "surveyors" operating in the 1820s included bed-ridden old men, coal merchants, bakers and publicans.[219] The situation was well summarised by John Clark (an associate of Arthur Young who we met in Chapter 3) who wrote in 1794:[220]

> The ridiculous farce of appointing one of the parishioners annually (at no salary) to enforce from his relations, friends, and neighbours, a strict performance of a duty which probably he never discharged himself; and from which, by shewing leniency to his neighbours, he will expect to be excused in his turn.

Despite its technical inefficiencies, the statute labour system was found useful in 18th- and 19th-century America. In Virginia, the people required to do a number of unpaid days a year working on the roads were called *labouring titheables*, and a 1785 law defined them as *all male persons over the age of sixteen, except ferry keepers and the master or owner of two or more labouring titheables.*[221] In 1829, the State of Georgia repealed its statute labour legislation and replaced it with a $70 000 allocation to purchase slaves to do the work.[222] The ancient Greeks had similarly used publicly owned slaves to build and maintain public roads. In 1904, 25 American States were still using the statute labour system, requiring all able-bodied males living along a road to work on its repair for a stated number of days a year, or to pay the equivalent in cash. In colonial America and Australia, convicts were often used for road-making and pavement repair.[223] In Alabama in 1913, the requirement was ten days labour or $5 cash.[224]

The system in Ireland in the 17th and 18th centuries was relatively well organised.[225] The main roads were the responsibility of each county and managed by a Grand Jury legally appointed by the High Sheriff of the County. The members were usually local land-owners, magistrates and sheriffs. It raised money by a form of land tax and had the required road-work done by contractors. The resulting roads were cheaper and better than their English equivalents.[226] A report in 1766 noted that *there was no country whatever where there were more or, in general, better roads than in Ireland.*[227] However, the view was not universally shared.[228]

An alternative to the various ways of taxing locals to maintain the nearby roads was to tax the road users when they used a piece of road or bridge. These taxes are called tolls, and the earliest recorded toll system was in India in 320 BC.[229] They were easy to collect at bridges and at gates in city walls but far less practical on open roads. Nevertheless, tolls were common throughout the 11th-century Europe and were specifically reported in England's Domesday Book in 1085.[230] Often tolls were seen by the local Lord as a means of taxation rather than as a source of funds to pay for road construction and maintenance.[231] The Great North Road out of London was subject to a number of unsuccessful attempts at tolling. Finally, an Act of 1661 gave tolling legal status and the first toll house was opened on the road at Wadesmill, about 35 km north of London. It was operated by local officials. Evasion was very high, and an Act in 1695 allowed barriers (toll gates) to be erected.

In 1706, Britain enacted the first of a long series of acts giving private trustees the power to collect tolls on lengths of roads that they had to (at least in part) construct and then fully maintain. They were called turnpikes as their gates were seen to resemble military weapons called pikes. There were over a thousand British turnpikes, each governed by its own parliamentary Act. After, the Revolution France used tolls to fund bridge construction.

Technical pavement issues associated with turnpikes were discussed in Chapter 3. Given that most road usage had been free, the turnpike system was never popular with toll-paying travellers and began to decline in the middle of the 19th century with the new railways siphoning off much of its more profitable business.[232] Rather than establish a national agency to manage the highways and major local roads, the British government reverted back to the failed earlier system of putting roads in the control of local agencies. This decision was administratively simple and aligned to the political opposition to centralised expert control then being advocated by the libertarian and free-market political movements.[233] Pavements again deteriorated, and the situation was only reversed by the irresistible pressures of the new motorcars at the beginning of the new century. Fuel tax became the new revenue source for building and maintaining roads and their pavements.

Chapter 8

Trésaguet and Telford lead
a pavement renaissance
in France and England

"Simple" pavements were discussed in Chapter 3, and in the post-Roman and medieval eras, the pavements being managed by the processes described in Chapter 7 were certainly still simple pavements. Progress in Europe required stronger national administrations that could see well-paved roads as an asset rather than as an invitation to outsiders to invade.

There were some signs of change in France late in the 14th century under Charles II, but the first precursive sign of a quantum shift in pavement technology occurred in late 15th-century France where Louis XI ordered construction of a new Royal Road between Paris and Orléans. It became known in the 16th century as the Queen of Roads and functioned as a toll road. Its pavement used 200×165 mm stone masonry blocks placed on a 300 mm thick bed of sand.

In 1508, Louis XII strengthened the regional road management system, in 1552, Henry II fixed the lines of the Royal Roads and, in 1555, he codified their pavement design.[234] However, the major change began in 1594, when Henry IV appointed a senior official (the Grand Voyer de France) to manage the national road system.[235] The position was abolished in 1627. Road development was given a further boost in the mid-17th century by the need to develop strategic roads to serve the Franco-Spanish War and French (Fronde) civil wars. Consequently, between 1300 and 1900, France probably had the world's leading road network with some 30 Mm of national ("Royal") roads.

In 1585, Guido Toglietta in Italy wrote a treatise on pavements in which he pointed out that lighter pavements than those used by the Romans would be feasible and effective. He emphasised using longitudinal drains, producing an impervious surface and enhancing a 50 mm thick pavement layer with the lime-based material.[236] Toglietta was probably referring to a cement-type mortar placed and pushed into the voids between the pavement stones. Toglietta's principles were very sound although his 50 mm mortared pavements would probably have been unable to resist the damaging effects of heavily laden wagons unless placed on a very strong and firm subgrade.

A detailed technical account of Roman roads was produced by Nicholas Bergier in 1622.[237] Bergier was a lawyer based in Rheims in France and criticised existing French roads *as a single layer of stones seated on ordinary sand without other support or foundation*.[238] Much of the detail of Roman roads in his book was based on a piece of Roman road within his family property. The book led to a lavish but technically unsuccessful revival of Roman road-making practice in France and in a number of German States. The basic cross-section used is shown in Figure 8.1.[239] The upper layer followed Roman practice and used broken stones rather the gravel.[240] A significant step occurred in 1672, when a powerful national figure, Jean Colbert, created the Corps of Military Engineers, with a major role in road building and maintaining. In 1693, Henri Gautier, who had originally trained as doctor of medicine, was appointed the national Inspector General of Bridges and Roads. In the same year, he published a book[241] on pavement construction which had a very elaborate pavement structure (Figure 8.2), reorienting some of the major stones to a vertical position and wedging the structure between kerb-like longitudinal retaining walls.

The use of stones smaller than 25 mm in the surface layer of Gautier's and all the subsequent unbound pavements to be discussed below was because they were smaller than the rims, hooves and feet of any expected road users. Thus, road usage would tend to compact the small stones into the underlying layers of larger stone although the presence of rounded stones in the thick surface layer would have made it difficult to maintain a good traffic surface.

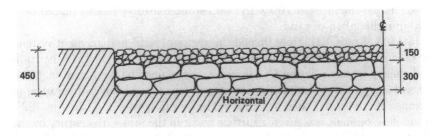

Figure 8.1 Pavement design used in the early 17th-century French revival of Roman road-making practice.[242]

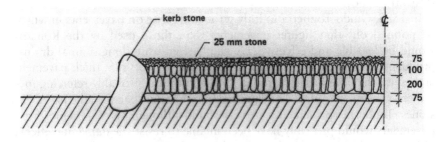

Figure 8.2 Pavement design used in Gautier's late 17th-century development of the earlier French revival of Roman road-making practice.[243]

On the other hand, if the stones at the surface were larger than the rims and hooves pressing on them, the stones would be dislodged or twisted as they were loaded eccentrically by the smaller rims and hooves.

An underlying assumption in both these two French methods is that the subgrade might not be adequate for its load-carrying task. Rather than improve the subgrade or use the Roman agger method, they concentrated more effort on the pavement structure. Gautier aligned two layers of stones vertically as they could be more securely placed vertically (horizontal stones would have required mortaring to prevent rocking under eccentric vertical loads), and as it was relatively simple to wedge smaller stones between them to aid horizontal stability between the kerb stones. A problem with the method was that it would not have helped distribute (and thus lower) contact pressures on the subgrade due to wheel loads and would have had some inherent lateral instability. The kerb arrangement shown in Figure 8.2 would have assisted the free-draining of the pavement structure.

More peaceful times early in the 18th century led, in 1716, to the creation of a central civilian road organisation, the Corps des Ponts et Chaussées, to act separately from the existing corps of military engineers and to be responsible for all major French road-making.[244] A significant advance on Gautier's method was introduced by Pierre-Marie Jerome Trésaguet in France in the latter half of the 18th century. Trésaguet was a trained civil engineer from an engineering family and had begun working on pavement issues in Paris in 1757. In 1775, A-R-J Turgot, Louis XVI's Finance Minister, appointed Trésaguet as Engineer-General of the Corps des Ponts et Chaussées. In the same year, Trésaguet presented a definitive memo on road-making to the Corps and it was later issued publicly.[245]

Trésaguet believed that the structure used by the Romans was unnecessarily massive and advocated the alternative pavement structure shown in Figure 8.3. His basic change from the Romans and from Gautier was that he relied on the subgrade to provide the pavement with its structural strength and stiffness. Consequently, he emphasised the Roman longitudinal drains and Gautier's kerb stones and required the subgrade to be compacted and then sloped to further aid the removal of water from the pavement. This attention to water was important as a major practical

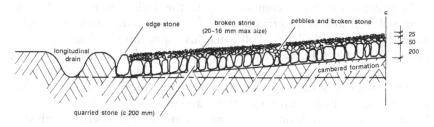

Figure 8.3 Pavement design due to Trésaguet in the second half of the 18th century.[246]

problem with Trésaguet's method was that the pavement was placed in a longitudinal trench that kept its surface at the level of the surrounding countryside and thus provided a longitudinal pond that unhelpfully retained water. The decision was "politically" wise as it did not compromise the drainage and accessibility of adjoining land-owners.

Trésaguet did not require Gautier's awkward bottom layer of horizontal stones, which he had found to settle unevenly, but he did use Gautier's layer of vertical quarried stones. This change means that Trésaguet required the subgrade surface to be of sufficient strength and stiffness to withstand the contact stresses at the base of the vertical stones. He alleviated this problem by requiring masons to ensure that the base surface of the vertical stones was relatively flat. However, if the subgrade was softened by water, the pressure from the base of the stones forced mud up between them and caused them to separate laterally.

Neither did he use Gautier's upper 100 mm layer but instead placed strong emphasis on the two upper layers making use of broken stone and avoiding rounded stones. He required these stones to be hammered into any gaps between the larger underlying vertical stones, leaving minimal voids. The surface layer of smaller stones was well-compacted but was also seen as a surface that could be readily maintained if damaged by traffic. In addition to being more effective in operation, the Trésaguet pavement was about half the thickness and therefore much cheaper than Gautier's earlier arrangement. However, the method spread only slowly through France and was subsequently quickly supplanted by McAdam's method (see Chapter 9), primarily because – relative to McAdam's method – the Trésaguet approach was more expensive and required skilled and committed workmen.[247]

Trésaguet advocated daily repair of the pavement surface using broken stones. To ensure that this continuous maintenance happened, he divided each road into longitudinal lengths to be repaired by a nominated paid workman (a *cantonnier*).[248] An English expert wrote very favourably of his observations of this method noting that the cantonniers *took a keen pride in keeping their stretch of road as smooth and clean as a billiard table.*[249] In the 18th century, Trésaguet's method was being widely copied in continental Europe and Russia.

As England developed commercially in the 18th century, the demand from political and commercial sectors for better roads with good pavements became irresistible.[250] Following his tours through England in the early 1720s, Daniel Defoe wrote:[251] *The inland trade of England has been greatly obstructed by the exceeding badness of the roads.* He had earlier noted[252] that one of the causes of this was an 80 km wide band of clay that stretched across England from sea to sea and that the "very great trade with London" had to pass through this band on roads that had been ploughed so deep that *the very whole country has not been able to repair them.*

There had been attempts from time to time to place the blame for pavement damage on wagons which were heavily laden and/or had wheel rims that were too narrow for the state of many roads. As traffic gradually increased, local attitudes to roads in the 18th century were well-described by the Webbs[253] who concluded that most 18th-century British road legislation was based on the:

> implicit assumption that the wheeled carriage was an intruder on the highway, a disturber of the existing order, a cause of damage – in short, an active nuisance to the roadway – to be suppressed in its most noxious forms and where inevitable to be regulated and restricted as much as possible.

Various measures were introduced to weigh wagons (by complex systems of levers) or to regulate the geometry of wheels. The measures had little or no success as the pavements were usually so poor as to be prone to damage by any vehicle. Furthermore, the wagons tended to self-regulate themselves as there was no merit to them in being bogged in transit.[254] This issue is pursued further in Chapter 19.

"Blind Jack" Metcalf, whose swamp-crossing work was referred to in Chapter 5 and who will be discussed further in Chapter 12, built roads in Yorkshire, Lancashire, Cheshire and Derbyshire in the north of England from 1750 to 1790. He was a strong proponent of surfacing a pavement with a compacted layer of small, broken stones with sharp edges. His general method was to remove the in situ material until he had reached a firm subgrade and then to dig longitudinal side trenches to ensure that the pavement would be well-drained. He next placed clean, broken stone on top of the subgrade to produce a convex road surface. His pavements were typically 300 mm thick with the size of the broken stones decreasing towards the pavement surface.[255] His method was very effective and was copied by many subsequent English road builders. It will be recalled from the earlier discussion of Figure 8.3 that in this same period Trésaguet was also advocating the use of layers of broken stone in French road construction.

A key response to the demands for better pavements came from Thomas Telford, a Scottish stonemason who had rapidly become a leader in the design and construction of buildings, bridges, roads and canals.[256] In 1803, Telford received a major and well-funded government commission[257] to provide roads as a consequence of rebellious Jacobite troubles in the largely inaccessible Scottish Highlands. He then proceeded to build over 1500 km of Highland roads during the next 25 years. Given his background as a stonemason and bridge-builder and his awareness of the remnants of Roman roads in Britain, it is not surprising that his roads drew significantly from stonemasonry and Roman practice. They were a great improvement on pre-existing English and Scottish roads but were also much more expensive than the earlier roads and this drew some trenchant criticism.[258]

The pavement specifications that Telford used "have not survived in full detail". However, the poet Robert Southey was a close friend of Telford and had toured Scotland in 1819 with Telford's associates.[259] In his diary of his tour, he describes the Telford pavement practice thus:[260]

> First to level and drain; and then to lay a solid pavement of hard stones, the round or broad end downwards; the points are then broken off, and a layer of stones broken to about the size of a walnut, laid over them so that the whole are bound together.

In a later specification[261] Telford added that *they are to be set on their broadest edges lengthwise across the road* (Figure 8.4).

The *level and drain* issue follows from our earlier discussion of Roman roads and assumes that the levelling produced a permanently firm subgrade. The layer of *hard stones* fills the same role as the statumen in Roman and Trésaguet pavements but was simpler to place as the need for close jointing between stones throughout their depth was eliminated by using a light hammer to place broken stones into the space between the hard stones. Telford also required the broken stones to be *as cubical as possible*[263] to avoid workers using long slivers of stone. Southey later gives "hen's egg" as an alternative definition of the maximum stone size (i.e. about 25 mm). The importance of this limit on stone size was that it was not greater than the width of the rims of wheeled vehicles or the hooves of animals using the pavement. This meant that under traffic, the stones would be pushed down vertically rather than twisted or pushed aside horizontally. When the stones did wear under traffic, the stone particles would helpfully fill the interstices between the broken stones, thus aiding the stiffness and reducing the porosity of the pavement. There are indications that this approach to stone size was based on the experience with the roads built by Metcalf in Yorkshire in the latter half of the previous century.

Aitken,[264] a fellow Scottish engineer, devotes about four pages of his book to what he says is Telford's "general specification for the Highland roads"

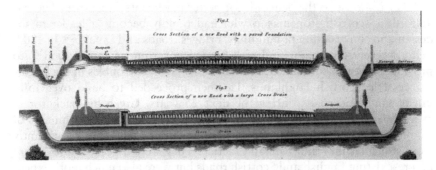

Figure 8.4 Telford's presentation of a cross-section of one of his roads.[262] The lower cross-section relates to his 1816 Glasgow to Carlisle road.

although his source is an 1847 publication.[265] However, by 1816 Telford was specifying a base layer of knapped stones *carefully set by hand ... and cross-bonded*.[266] With his bridge and masonry background, Telford explained his bottom layer by noting that:

> roads are structures that have to sustain great weights, and violent percussion, the same rules therefore ought to be followed in regard to them as are followed in regard to other structures.[267]

A typical Telford pavement cross-section is shown in Figure 8.5. Aitken's interpretation of Telford's pavement is shown in Figure 8.6.

Another important restriction due to Telford (and probably Metcalf) was to eliminate the use of clay and fine soft rock such as chalk to fill voids in the broken stone layer. This had been a common technique used to provide a smooth and bound surface on a newly made pavement. It lasted as long as the first heavy rain when the filler turned to a viscous fluid which served mainly to aid the displacement of the pavement stones.

Telford was aware of Trésaguet's approach and some of its associated problems and consequently he used larger and more cubical stone blocks in his base layer. Their longest edge was placed transverse to the traffic direction and each transverse line of blocks was staggered in the manner of a conventional block (or brick) wall. The smallest faces of the blocks were their upper and lower surfaces. This meant that the side faces of the blocks were much closer to vertical than in Trésaguet's method. Finally, the top edges of

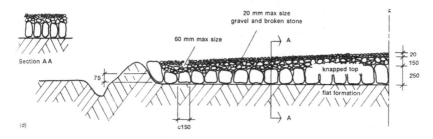

Figure 8.5 Telford's cross-section.[268]

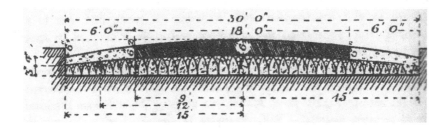

Figure 8.6 Aitken's 1900 representation of a Telford pavement.[269]

the blocks were knapped to make it easier to wedge smaller stones between them. All this meant that placing lowest course of the pavement required much more work and was more expensive than was Trésaguet's method.

Note in the diagrammatic cross-section shown in Figures 8.4–8.6 that Telford reverts back to the horizontal subgrade of Gautier (Figure 8.1) and drains the subgrade surface by lateral drain-pipes that can be seen in Figure 8.4, rather than using Trésaguet's transversely sloping subgrade. Where feasible, Telford raised the level of the top of the subgrade above the level of the adjacent countryside. Telford was very focussed on the quality and extent of the drains[270] and, for instance, required the base of the drains to be paved where the soil was soft or loose. The major effort that Telford devoted to keeping water away from his pavements and subgrades was in strong contrast to many generations of his road building predecessors.

While we might well consider Telford's pavements, in hind-sight, to be over-designed, they were a major advance on previous pavements and served their purpose well, particularly on his Holyhead (or Great Irish) Road which was one of the 19th century's great pieces of road design and construction. Whereas the pavements he used were based on traditional masonry concepts, his road alignments were based on innovative tests by Telford and his colleague John Macneill.[271]

Chapter 9

McAdam invents a major new pavement using broken stones

Although Telford's method described in Chapter 8 was very effective, it was fortunate that its use was funded by major government agencies such as the Postmaster General[272] as it was too expensive for many more mundane road locations. The need for a cheaper but still effective method of pavement construction heightened as traffic increased as a consequence of the Industrial Revolution and as toll roads (or turnpikes) became more widespread.[273] Into this market void stepped John McAdam, a businessman and property manager who subsequently also became a turnpike manager (Figure 9.1).[274] McAdam was also born in Scotland and within a year of Telford.[275] He had been interested in roads as a boy but had worked as a merchant in New York from 1770 to 1783 and had married a woman from a wealthy New York family.[276] He returned to Scotland after the American War of Independence. Some personal aspects of his life are discussed further in Chapter 12.

Figure 9.1 John Loudon McAdam.[277]

In about 1802, he began maintaining and then building roads in the Bristol area of south-western England. After observing the performance of many existing pavements, he came to the conclusion that a well-compacted layer of broken stones placed directly on a firm subgrade could be effective without the need for the statumen layer of larger stones used by the Romans, the French and then by Telford. Addison claims without offering supporting evidence that the method had previously been used in Switzerland, Sweden and "other countries".[278] However, other evidence[279] suggests that reference to Sweden and Switzerland was only to the use of broken rather than rounded stones. To assist removing water from the pavement, McAdam followed Trésaguet rather than Telford and sloped his subgrades across the roadway towards roadside drains (Figure 9.2). We will return to the vexed question of the subgrade and its properties later in the chapter.

McAdam often obtained his broken stones by employing local labour to work at the roadside breaking down larger stones left behind by the previous attempts at road-making. For example, the London Times on 8 October 1824 described the macadamising of Whitehall St in London thus:

> The great granite stones are broke into small pieces as soon as they are taken up, and thus, as rapidly as the way is cleared the materials are ready for the commencement of the Macadamizing system.

McAdam made some quite specific comments on the stone-breaking process:[281] *The only proper method of breaking stones, both for effect and economy, is by women, boys or old men past hard labour, sitting with small hammers of about 500 gm weight with a short handle.* McAdam's alternative to the walnut and hen's egg definitions of stone size employed by Telford was that the broken stone used in the upper 50 mm layer pavement had to be capable of fitting into the local stone-breaker's mouth. The story is told of McAdam accusing a workman of placing stones that were too large. The workman's denial was based on displaying *a mouth of extraordinary capacity and totally devoid of teeth.*[282] In today's terminology,[283] the use of stones of a single size would be called *open grading*.

If more precision in stone size was needed, Telford specified use of a ring gauge and McAdam required use of a pocket balance and a 170 gm

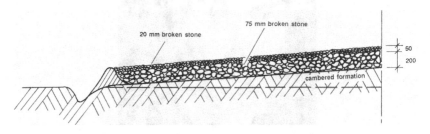

Figure 9.2 McAdam's cross-section in the style of Figures 8.5–8.7.[280]

weight. McAdam noted[284] that *a woman sitting will break more limestone than two strong labourers on their feet.* Telford similarly noted that such stone-breaking was not onerous work and could be done by *men who are past hard labour, and women and boys.* He also suggested that the work-men shovelled the broken stones with rakes whose prongs were too closely spaced to allow the passage of large stones.[285] A good stone-breaker would produce about 4 m³ of stone per 12 hour day. Hand tools in use at the time are shown in Figure 9.3. Stone-breaking using such tools came to be widely

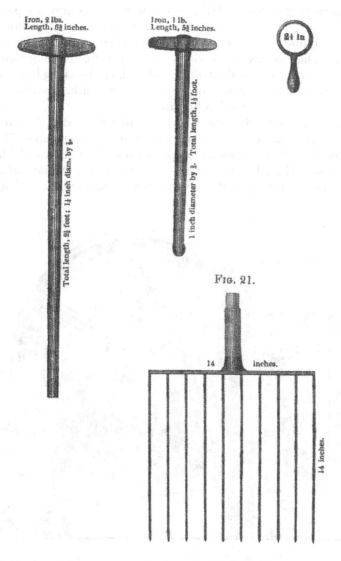

Iron, 2 lbs.
Length, 6¾ inches.

Iron, 1 lb.
Length, 5¾ inches.

2¼ in

Total length, 2¾ feet; 1¼ inch diam. by ⅞.

1 inch diameter by ¾. Total length, 1½ foot.

FIG. 21.

14   inches.

14 inches.

Figure 9.3 Hand tools for pavement maintenance in the 1840s.[287]

used to provide work for the unemployed and for convicts. For example, in the late 1860s over 400 men were being given unemployment relief, breaking stones with hand tools while seated under the railway arches at Bethnal Green in London.[286]

Steam-powered stone-breaking machines were first produced by Eli Blake in Connecticut in the 1850s. Blake was a lawyer on a committee overseeing the construction of a macadam pavement in New Haven. He had been appalled by the two days' labour that took a man to crush a cubic metre of stone by hand and developed steam-driven mechanical stone crushers called jaw crushers (Figure 9.4).[288] His crushers were used to produce pavements for Central Park in New York. Other crushing devices used gyratory, hammering or rolling actions. Initially, stones broken by machines were used with some reluctance as it was widely believed (1) that hand-broken stones produced better mechanical interlock and (2) hand-breaking had become a major source of local employment. By 1862, stone crushers built to an 1858 Blake patent were being used in Britain.

As shown in Figure 9.2, the subgrade is first covered by a lower stone layer about 200 mm thick and comprised of broken stones up to 75 mm in size. This statumen-less, non-Telford solution worked because the layer of broken stones behaved as a coherent structural layer spreading any applied loads so that the stresses on the subgrade were reduced to a level that would

Figure 9.4 Blake's oscillatory stone crusher.[289]

not cause it distress. This use of larger stones in the lowest pavement layer was an indirect recognition of Telford's approach.

McAdam objected to the stones being used in a pavement layer having a range of sizes and minimal air voids (a well-graded mix, Chapter 4) as he thought this would give pavement makers an opportunity to surreptitiously use such unsuitable materials such as round sand, clay, mud and miscellaneous surface wastes for the smaller-sized components. The main problem with clay and other soils was that they were washed away in wet weather and were blown away in dry and windy weather. Later, the availability of sound fine stone as a by-product of steam-powered stone crushing (Figure 9.4) removed the basis for objecting to well-graded mixes.

It has been suggested[290] that McAdam's move away from a block-based statumen might have been prompted by a lack of suitable stones for block-making in the parts of England where he began his major involvement in pavements. However, the suggestion came from a Glaswegian who claimed no direct experience with English roads. There is little doubt that McAdam's move was motivated by cost rather the availability of suitable stones. Many claimed[291] that McAdam based his method on the work of Phillips (Chapter 3) but the two approaches are totally different. The confusion might have arisen from Phillips' advocacy of the use of uncompacted "sharp stones", but by this Phillips meant loosely placed washed (rounded) gravel[292] and certainly not the interlocked broken stone.

McAdam required his broken stones be placed in layers about 50 mm thick and for each layer to be allowed to consolidate under traffic. Any ruts were then filled with new broken stones before the next layer was placed.[293] McAdam argued that the attrition and compaction of properly sized stones would be sufficient to produce a pavement with adequate strength and stiffness. Any subsequent irregularities or weakness in the subgrade could be quickly remedied by placing further stones.

The broken-stone principle is still used in modern highway construction,[294] and since 1820, McAdam's name has been remembered by the term *macadam* used to describe courses of compacted broken stones working together due to natural interlock between the stones. The associated method was called *macadamising*. However, in some places the broken stone is referred to as *road metal* and resulting pavement is described as a *metalled road*. The use of the word *metal* in this context came from Latin *metallum*, meaning a mine or quarry. Thus, *metal* originally referred to anything useful that had been extracted from the ground.

McAdam was not the first to use this simple and practical approach as earlier chapters have mentioned the use of stones rammed into subgrades.[295] Nevertheless, the total reliance on broken stones must have appeared to many early pavement makers as too obvious to be successful. What happens in practice – particularly with McAdam's approach – is that the compacted layer of broken stones behaves as a single coherent layer and not as a collection of individual stones. Thus, surface loads are well distributed by structural interlock and by stone-to-stone contact before they reach the subgrade, avoiding the need for the masonry-like (and expensive) layer of large stones used by Gautier, Trésaguet and Telford. This was achieved by preparing the layer of broken stones *to unite by its own angles into a firm, compact, impenetrable body.*[296]

As will be discussed in later chapters, there was also a growing push to apply various *glues* (called *binders* in the pavement world) to pavement surfaces. McAdam saw interlocked stones as a far better alternative and wrote *nothing is to be laid on the clean stone on pretence of **binding**; broken stone will combine by its own angles into a smooth solid surface that cannot be affected by the vicissitudes of weather, or displaced by the action of wheels.*[297] To protect the subgrade from water, McAdam also required the pavement surface layers to be *perfectly impervious to water.* The *smooth solid surface that cannot be affected by the vicissitudes of weather* was achieved by the use of smaller stone pieces from the stone-braking process which were hand-rammed into surface voids. McAdam did not restrict the use of smaller broken stones or sharp sand. Rather, he clearly prohibited the use of *earth, clay, chalk, or any other matter that will imbibe water, and be affected with frost; nothing is to be laid on the clean stones with the pretence of binding.*[298]

Then, when the pavement was being trafficked, the action of the hard iron-shod wheels and hooves of the initial traffic would cause grinding of the stones at or near the surface and would then push this by-product into any remaining surface voids. McAdam required his stones to be under 20 mm in size at the surface of the pavement as this would be smaller than the size of those wheels and hooves which could not, therefore, tilt the stones by landing only on an edge. As McAdam said, [299] any larger stones would *be in constant motion* under traffic.

Later, we will discuss how the use of cementitious and bituminous mortars to play this space-filling role produced a whole new generation of road pavements. In this context, McAdam believed that attrition due to passing traffic would adequately fill any interstices in the surface layer. Where this was not adequate, a process known as *blinding the open eyes* was used. This involved sweeping chips from the stone-breaking or sharp-edge sand into any remaining surface voids. The resulting pavement was called *dry-bound* or *traffic bound*. Another version of this approach is called *water-bound macadam* and relies on the rather fleeting cementing powers due to surface tension of any water wetting the fine material within the interstices of the broken stone layer. When steam-powered stone crushers became available in the second half of the 19th century (Figure 9.4), sound fine stone was readily available as a by-product of the crushing process.

After opening the road to traffic McAdam required *a careful person to attend for some time after the road is opened for use, to rake in the track made by wheels.*[300] As mentioned in Chapter 8, Trésaguet had also successfully used this method. Telford and McAdam's pavements did require regular maintenance. When macadam was under heavy traffic, the surface was cleaned and raked on a daily basis and any dislodged surface stones were replaced. Iron-rimmed wheels and iron horse-shoes readily ground stones in the surface to a fine powder. Even under light traffic, about every 5 years the whole surface had to be *lifted*, a process in which picks were used to merge new stones into the top 50 mm of the pavement. In 1823, Telford wrote *A certain number of labourers should always have care of the surface of a road and never quit for a single day.*[301] With the coming of the railways in the 1830s, it became increasingly difficult for turnpike companies to find the revenue to undertake such tasks, and many turnpike roads sank into disrepair.[302]

The improvements produced by the work of McAdam and Telford did have noticeable effects. Young's comment on the use of large boulders in Oxfordshire road-making in the 1770s was quoted earlier. In 1809, however, he was able to observe that in the same Oxfordshire *roads are greatly improved ... the turnpikes are very good and where gravel is to be had excellent.*[303]

A recent misinterpretation of McAdam's approach occurred in 1953 when RJ Forbes, the author of a number of otherwise excellent road histories, wrote[304] that McAdam's *system of well-rolled layers of stones with sand and loam washed into the interstices was not an original invention. It had been used in Sweden and Switzerland long before.* But this was not McAdam's invention – as we have seen McAdam objected to the use of loam and had realised that a layer of broken stones could be made to act as an independent structural pavement. A more perceptive observer of Swedish road-making in the 1820s had commented[305] *Good rock is generally met with in Sweden; and they spare no pains in breaking it small.* Sweden at the time had an international reputation for the quality of its roads and pavements.[306]

There was much public and professional debate about the relative merits of the methods of Telford and McAdam. It reached a peak in the second quarter of the 19th century but was still continuing a century later. The basis of Telford's method was obvious, but it was relatively expensive. On the other hand, the basis for McAdam's method was not immediately obvious and this made many potential users sceptical. In its favour, it was cheaper and performed surprisingly well under traffic. At the beginning of the 20th century, the Telford base layer of knapped stones (Figure 8.5) continued to appeal to many pavement builders. For instance, it was still widely used in the eastern parts of the USA and in Switzerland.[307]

McAdam's method had gained rapid acceptance in England, helped by his own tireless advocacy and the wide distribution of his paving manual.[308,309] Using these guides, the first McAdam pavement was placed in Australia in 1822 between Prospect and Richmond in NSW[310] and in America in 1823 on a turnpike road between Hagerstown and Boonsboro in Maryland.[311] A leading French civil engineer, Charles-Louis Navier,[312] visited England in 1823 and returned to France, enthusiastically supporting McAdam's approach. Official blessing occurred in 1830 and by 1843 it had been widely but gradually deployed in France.[313]

Pavements built to McAdam's principles were soon being called *mac-adam* pavements. The term originally implied a mix of broken stones, each less than 30 mm in size. Nevertheless, the name *macadam* was soon being misleadingly applied to a range of stone pavements. By the end of the 19th century, there were sound and comprehensive manuals available on all aspects of macadam pavements.[314]

Edward North was a well-known American road builder who wrote[315] in 1879 that McAdam had systematised the construction and maintenance of roads using *small, clean stone* without mentioning that the stones should be broken. Indeed, he later commented[316] that *In Paris the macadam roads are composed of water worn flint pebbles.* At the very least, this indicates that the term *macadam* was being loosely used to cover a wide range of pavements. North's own descriptions are ambivalent for he later writes *both the English and French prefer hand broken stones for macadam* rather than machine-broken ones.[317]

The common-sense adoption of McAdam's method is perhaps best reflected in an 1850 publication[318] on constructing and repairing common roads by Henry Law, a well-regarded English civil engineer who had trained under Isambard Brunel. Law wrote (p104):

> The method of forming the road covering entirely with angular pieces of stone, without any other material, was first strongly recommended by Mr McAdam, and all subsequent experience has shown its superiority over every other method that has been employed.

Law was correct, and McAdam's methods have been widely applied from 1836 to the present day. Joseph Needham in his 1971 epic study[319] of China incorrectly described Chinese pavements as water-bound macadam produced from compacted rubble and gravel. In 1979, Clay McShane,[320] an otherwise respected commentator on asphalt roads, described macadam as *a technically superior method of using gravel.*

Japan's famed Tokaido Road had developed during the 17th and 18th centuries (today it links Tokyo and Kyoto). In 1859, the first British Ambassador to Japan wrote glowingly of the road:

> The Tokaido may challenge comparisons with the finest in Europe. The roads are broad, level, carefully kept and well macadamised, with magnificent avenues of timber to give shade. It is difficult to describe all their merits.[321]

However, many consistently misinterpreted McAdam's approach, perhaps because it was counter to their intuitions or to their current practices. Sir Henry Parnell (Baron Congleton) was an eminent figure in English politics and engineering and a close associate of Telford. In his widely read book, he struggles to not specifically mention McAdam and writes, for instance, that:

The business of road-making in this country has been confined almost entirely to the management of individuals wholly ignorant of the scientific principles on which it depends. Nothing can be more opposed to the principles of science than what is recommended (in McAdam's work).[322]

In an otherwise thorough review of English turnpike roads of the time, another expert failed to grasp the significance of using broken stones in road-making and repair and focussed more on McAdams' "elastic" road surfaces – a concept alien to McAdam but probably based on Parnell's pointed and somewhat bitter criticisms of McAdam's methods.[323]

Another contemporary objection came from James Paterson,[324] a competent road surveyor from Montrose in Scotland who was a strong advocate of the use of broken stones. He petitioned the British parliament in 1825, claiming that McAdam's invention was not original as he and others had previously used it. Both McAdam and Paterson advocated the use for broken stones, but Paterson did not appear to grasp the importance of an interlocked compacted surface layer. Two key differences between Paterson and McAdam were that Paterson advocated that: (1) the bottom layer should be composed of *stones broken large and, of course, the softer the bottom the larger (*up to 75 mm*) they should be*, and (2) *the materials on the top should be bound together by throwing on a little earth or small sharp gravel*.[325] However, McAdam's basic concepts opposed the use of *a little earth* as a surface binder and did not accept "soft" subgrades.

In Paterson's defence, it should be noted that both he and McAdam were at the forefront of the move to keep pavements and subgrades dry, particularly by the use of longitudinal drains and by raising the pavement above the level of the surrounding land. This was particularly important when a macadam pavement was placed over a moisture-sensitive subgrade such as one containing clay. In his book, McAdam was very clear on these issues. In particular, to manage water from the adjacent ground, he wrote[326] *The first operation in making a road should be the reverse of digging a trench. The road should not be sunk below but raised above the level of the adjacent ground.*

Nevertheless, in 1876 when Robert Moore presented what was described as an *exhaustive review of street pavements*[327] when referring to the debate between the methods of McAdam and Telford when applied to subgrades he claimed that (p146):

there is no longer any controversy. All road builders who have given the subject any study admit that the Telford system is the best. The large stones of the Telford subpavement distribute the load over a large surface and do not sink into the earth as the smaller stones will do.

The Scottish misinterpretations continued as in 1878 William Lochhead claimed[328] that McAdam had copied the work of his father on the

Glasgow-Paisley road. McAdam may well have observed Lochhead Senior's emphasis on stone size, but Lochhead had also followed Telford in using a bottom stone layer of 200 mm freestone stone blocks set on end. Freestone is commonly a fine-grained stone that can be readily cut by stonemasons.

As recently as 1965, Desmond Clarke, a well-known Irish professor of philosophy, wrote[329] astonishingly,

> the methods McAdam adopted and advocated were incorrect, and in fact were not applied in road building from the date of his death (1836). Edgeworth and McAdam followed the same pattern of road construction up to a point, that is a foundation of large flat stones upon which were laid stones of six or seven pounds weight, with a final covering of small stones an inch or two in size.

Clarke's second sentence is perversely wrong as the method he describes is a reasonable interpretation of Edgeworth's quite sensible approach[330] which was much more like Telford's method (Figure 8.5) than McAdam's (Figure 9.2). Edgeworth emphasised the need to use broken stones but used Telford's bottom layer. The avoidance of the bottom layer of large flat stones was one of McAdam's major innovations. On the following page, Clarke also misinterprets McAdam's somewhat obscure comments on binders, which we have untangled above.

Anderson was one of the few non-technical authors to grasp McAdam's broken stone concept. In 1932, she wrote[331]:

> macadam was finely broken stones that would unite by their own angles, and instead of using binding material such as dirt, chalk, or clay, he filled the spaces between the layers of broken stone with much more finely broken angular stones so that in wearing, instead of loosening as the surface had always done hitherto, the passage of wheels and the action of horses' hooves would simply bind the surface together.

Over time, roads made from broken stones were often a compromise between the methods of Telford and McAdam. However, the Telford bottom layer of carefully placed large stone blocks was rarely used but was replaced by a compacted layer, perhaps 150 mm thick, of much larger broken stones. Given the many related "criticisms" of McAdam's work discussed above, the placement of layers of broken stones and the underlying subgrade both deserve more discussion. With respect to the layers of broken stones, it would have been obvious to an observer that such layers rarely settled to a stable configuration without some extra help, most commonly by being compacted by some applied vertical loads. The compaction would continue until no further vertical deformation occurred, a situation known as *refusal*.

If layers were not compacted to refusal by their builders, they soon would be compacted by passing traffic and the resulting uneven surface would demand attention, as McAdam had foretold.

The stumps were sometimes called *tampers* or *pelons* in French. Ramming, or tamping, to refusal was a well-practised technique in pavement construction. Chinese road builders were using metal rammers during their peak road building period in about 200 BC.[332] A frieze in the Museum of Roman Civilisation in Rome shows workers in the 1st century AD using timber stumps about 1.5 m long. Later workers used large stumps as rammers, and in the 18th century, a range of wooden and iron-reinforced hand rammers were being manufactured by E Krantz in Breslau.[333] The bottom left corner of Claude Vernet's famous 1774 "Construction of a road" painting in the Louvre shows on the lower left a large manually operated compaction tool in use. By the 19th century, European rammers weighed about 20 kg and had iron bases. The 1851 lithograph in Figure 9.5 shows workers are compacting in time to a rhythm set by the foreman-cum-conductor on the right. An American, Ross, developed a steam-powered rammer in the 1860s (Figure 9.6).

The compaction process was both improved and quickened by the use of rollers. These had been employed by the Romans (Chapter 6), and a 1725 German book[335] shows a horse-drawn cast-iron roller. German cast-iron rollers were over a metre in both width and diameter. They applied loads of over 5 ton and were drawn by *six strong horses*.[336] Rolling then became an official requirement in the construction of German (Prussian) roads.[337] Such rollers were subsequently introduced into France in 1787 by Louis-Alexandre de Cessart,[338] Inspector General des Ponts et Chaussées.

Figure 9.5 Paving blocks being compacted into place on the Strand in London in 1851, near Temple Bar. The blocks are supported by a concrete slab (see Chapter 10).[334]

Figure 9.6 Ross' steam-powered road rammer. It weighed about 20 ton. Nothing is known of its applications.[341]

In 1817, Phillip Clay in England developed a horse-drawn roller which could apply a load of up to 20 ton.[339] The British had been initially concerned with the possible damaging effects of heavy rollers. However, John Burgoyne, a British Army engineer working in Ireland in the 1820s, presented careful and well-received observations as to how they might be used.[340] Burgoyne had served with distinction in France and was later placed in charge of the Irish transport system, where he made exceptional improvements. The success of the Irish experience meant that heavy rollers became more common late in the 1820s.

Rolling was sometimes not very effective, and McAdam had preferred compaction by traffic rather than by rolling.[342] Writing in 1834 one English pavement expert commented[343] *even with the best gravel and rolling, worms still find their way through and bring earth to the surface.* Nevertheless, within a few years of their introduction the advantages of rolling a pavement of broken stones with heavy rollers became manifest.

In particular, rolling:

1. detected weaknesses in a pavement prior to opening it to traffic,
2. achieved maximum practical compaction of the interlocking stones, reducing future settlement,

3. helped fill the surface interstices in the pavement, providing a surface which was smoother and more waterproof,
4. provided an even overall surface, and
5. dramatically reduced wear of the macadam surface by minimising relative movement of the surface stones.[344]

Thus, heavy rollers came to be seen as an essential tool in the production and maintenance of high-quality macadam pavements.[345] With asphalt paving (Chapter 13), the additional value of rolling was that it made it possible to produce a smooth surface.

There was a limit to the weight of roller that a horse could pull, and it was evident that that weight was insufficient to properly finish a macadam pavement.[346] A further problem with horse-drawn rollers was that the horses' hooves often damaged the pavement surface as the horses struggled find sufficient purchase to draw the heavy rollers. However, the need for change was driven mainly by the growing use of asphalt paving, to be discussed in Chapter 13, and steam-powered road vehicles were becoming common in the 1840s.[347] The French began using steam-powered rollers in about 1850 during trials of asphalt on the Avenue de Marigny in Paris. The device was probably designed and built by Louis Lemoine of Bordeaux who took out a patent in 1859 and used his machine in Bordeaux and in the Bois de Boulogne in Paris.[348] The Paris Steam Road Rolling Company then produced a steam roller patented by M Ballaison which was also used in Paris in 1862. The Paris municipality conducted comparative trial of horse-drawn and steam-powered rollers. This led, in 1865, to the municipality giving a 6-year contract to the Paris Steam Road Rolling Company to provide seven steam rollers for the continuous use of the municipality.[349] A French road roller that could be drawn by horse or steam-power is shown in Figure 9.7. The use of rolling increased dramatically following the introduction of steam-powered rollers.

Louis W Clark in Kolkata (Calcutta) was dismayed by the problems he was having with road rollers drawn by bullocks[351] and, in 1855, he asked W Bathe in Birmingham to produce a steam-powered roller.[352] It was sent

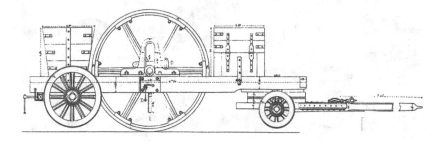

Figure 9.7 French road roller, c1860. The drum was cast-iron and contained ballast (from an 1860s French technical magazine).[350]

to Kolkata in 1864. In 1865, Thomas Aveling in Rochester in Kent became involved; he used one of his steam tractors to pull a separate roller in the manner of Figure 9.7. In that year, a 20 ton Aveling machine was successfully used in a road-making trial in Hyde Park in London. The construction advantage of these new rollers was that they produced the same amount of compaction in a day as would previously had occurred after at least three months of real traffic, thus further obviating much of the need for ongoing resurfacing of macadam pavements.[353]

By 1867, Aveling was producing rollers combined with a steam engine (Figure 9.8). These were far more manoeuvrable than the hauled rollers and became widely used. It was subsequently found that increasing the roller mass to 30 ton tended to excessively crush the stones being compacted.[354] The introduction of internal combustion engines at the turn of the 20th century[355] allowed the development of rollers that were even more manoeuvrable and more environmentally suited to work in urban areas (Figure 9.9). The subsequent use of the vibrating rollers to be discussed in Chapter 15 was to provide further beneficial outcomes.

Steam-powered rolling also led to major improvements in unbound macadam pavements as each layer of broken stones could be methodically rolled dry of water to "refusal", often indicated by a mosaic appearance of the rolled surface. Small particles of stones were then used to fill any joints in the mosaic. An observer described[357] the process in Liverpool in 1867 when the roller shown in Figure 9.9 was applied to macadam pavements: *The road is soon brought down to an even surface by the roller. The little crevices*

Figure 9.8 The first production steam roller made by Thomas Aveling for the Liverpool Corporation in 1867.[356]

Figure 9.9 An early Barford petrol-powered road roller.[358]

that have been made are then filled in with finer stone and again rolled,
until the surface of the whole street is perfectly firm, smooth and consoli-
dated. The process also avoided the harm being done to horses when they
fell on the exposed points of broken stones at the surface of unconsolidated
macadam.

The use of pavement layers comprised of a single layer of stones of a
single size had a number of drawbacks. First, it made the layer very porous
permitting water to pass through to the subgrade, which could be moisture
sensitive. Second, the interstices tended to fill up with rubbish, particularly
at the surface. With the introduction of powered pavement rollers, it was
found that using a mix of stone sizes such that the smaller stones filled all
the spaces between the larger stones – a *well-graded* mix – not only made
the layer immediately impermeable without waiting for McAdam's attrition
processes discussed above but also aided its compaction and the ease of
providing a level surface. Even with modern construction equipment, there
is a practical limit to the thickness of stones that can be properly placed in
one pass of the equipment.

With respect to the underlying subgrade, it is useful to quote McAdam's own words:[359]

> The following principles (must) be fully understood, admitted and acted upon: namely, that it is the native soil which really supports the weight of traffic: that while it is preserved in a dry state, it will carry any weight without sinking, and that it does in fact carry the road (i.e. the pavement) and the carriages also; that this native soil must previously be made quite dry, and a covering impenetrable to rain must then be placed over it, to preserve it in that dry state; that the thickness of the road should only be regulated by the quantity of material needed to form such impervious covering, and never by reference to its own power.

In addressing an audience of turnpike trustees, he wrote:

> Were the natural earth always dry, nothing could be preferable for being travelled upon, it would never wear out, nor would any carriage, however heavy, sink into it. The object to be aimed at, therefore, is to keep the natural soil dry, by defending it from groundwater and from that which falls from above. In the knowledge of the measures requisite to effect these objectives, consists the whole science of road building.[360]

Thus, McAdam's elimination of the lower basecourse of large carefully placed stones used by Telford, and most earlier road makers must be tempered by the fact that he only placed his pavements on firm subgrades and in many cases on pre-existing pavements. If the subgrades were not firm and could not withstand the compaction of upper layers of the basecourse or the loads from subsequent traffic, then the various enhancements used by the earlier road makers were an inevitable necessity. Most of McAdam's roads were enhancements of existing roads so the subgrade properties would have been well-known. The McAdam approach is clearly not directly applicable to swampy or even boggy ground.

In conjunction with all this, McAdam also required the pavement raised above the level of the adjacent ground and longitudinal drains to be *some inches* below the level of the ground.[361] Like Telford and the Romans, McAdam required the lowest point of his pavement to be about 100 mm above the maximum water level in the longitudinal roadside ditches. This allowed any water that had entered the pavement to drain from the edges of the pavement to the ditches.

Forbes extended his earlier misunderstandings when, in 1958, he wrote *nowadays [McAdam's] idea of a soft, yielding foundation for the road is not accepted as good practice*.[362] This view had previously been widely promulgated by Hartman who wrote[363] in his 1927 book *The flaw in McAdam's method was that he provided no foundation for his metalled surface*. McAdam's subgrades were not necessarily as strong and stiff as Telford's, but he believed that by keeping them dry and removing unsuitable material they would be more than adequate. The outspoken McAdam did not help in this part of the debate as, to contrast his approach to Telford's, he publicly and provocatively maintained his preference for a so-called yielding and soft foundation over one which was firm and unyielding.[364]

It must be said that many of the strictures of Telford and, particularly, of McAdam were made irrelevant by the widespread availability of heavy steam rollers. For example, it was no longer necessary to wait for the pavement to consolidate under real traffic and to repeatedly remedy the surface effects of that consolidation. Similarly, at one extreme stones larger than the hen's egg could now be effectively consolidated, and at the other extreme, fine particles could be added to the initial mix as they would aid roller compaction. There was no longer the need to wait for them to be produced by the passage of traffic.

As fast-moving pneumatic-tyred vehicles appeared on roads in the 20th century, [365] a new set of problems arose as the aerodynamics of the speeding vehicles produced near vacuums at the pavement surface and strong air currents across it. Dust clouds were instantaneously formed. Whereas the iron-tyred vehicles had produced stone dust to fill interstices in the surface, the new generation of speeding vehicles soon removed most of the fine surface material and left the pavement both porous and rough. The airborne dust had been part of the grading of the pavement basecourse and its absence created surface ruts and voids within the basecourse, reduced particle interlock and increased permeability. By 1907, LW Page, Director of the US Office of Public Roads, wrote:[366]

> The existence of our macadam roads depends upon the retention of the road-dust formed by the wearing of the surface. But the action of rubber-tire motor-cars moving at high speed soon strips the macadam road of all fine material, the result being that the road soon disintegrates.

As traffic increased, the cost of maintenance also increased, and in urban areas, the daily cost of removing dust in dry weather and unsanitary mud in wet weather made macadam pavements increasingly unattractive. By the mid-1930s, many existing macadam pavements had been surface dressed with bitumen or tar (Chapter 14) and, due to their well-tested subgrades, they often gave many years of further service.[367] Furthermore, the more coherent pavements discussed in the following chapters were beginning to provide realistic alternatives to macadam. By the mid-20th century, water-bound macadam was only being used for very lightly trafficked pavements.

The road dust also caused a great public nuisance, as Rudolf Diesel,[368] who was an avid long-distance motorist, describes in his 1905 diary. He was travelling down an Italian valley and looked back to see that his car had filled the whole valley with a wondrous white cloud of swirling dust:[369]

> What a dust storm we stirred up leaving Italy! Powdery limestone dust lay 50 mm thick on the street. Georg raced along, demanding from

the car all it had to give, through the Piave Valley, and behind us there swelled a colossal cone. This white cone rose in the air and expanded to infinite proportions. The entire Piave Valley was thick with the fog, a white cloud lying over the valley all the way to the mountain ridge. We outraged the pedestrians with gas attack – their faces pulled into a single grimace – and we left them behind in a world without definition, in which the fields and the trees in the distance had all lost their colour to a dry layer of powder.

Motorists in the early open cars wore protective clothing to counter the adverse effects of dust.

With all the above concerns about road dust at the beginning of the 20th century, it is not surprising that when the first meeting of the international association of road agencies (PIARC – Permanent International Association of Road Congresses[c]) held its first Congress in Paris in 1908, the third "question" on its agenda was "How to reduce wear and dust". The chairman of the new organisation noted that dust was *making journeys unbearable for road users and life impossible for people living along roads.*[370]

The dust was not only a major annoyance to travellers, to pavements, to people and property adjacent to the road but was also widely believed to be a health hazard to the lungs and to transmit life-threatening diseases such as tuberculosis.[371] It certainly killed roadside vegetation and depressed land values. Consequently, the public pressures for coherent (*sealed*) road surfaces was very strong but proved particularly difficult to address with respect to the long lengths of rural roads that existed in sparsely settled areas outside cities and towns. Water carts were used to wet the road surface and provide temporary relief by holding the dust particles together by capillary action.

Many dust palliatives were marketed with names such as Antistoff, Dustabato, Pulvercide and Tarraolio.[372] They worked by binding the dust particles together and therefore making them less likely to become airborne. They rarely lasted beyond the next rain shower. An American technique which reportedly lasted for about six months was to collect straw after grain harvesting and mix the chopped straw with clay. The moistened mix was then applied to the road surface.[373] One method tried by the American Office of Public Roads in 1908 was to coat the pavement surface with a mixture of molasses and quicklime. It was sticky enough, but quickly dissolved in water.[374] Molasses had been tried previously in 1895 in Chino, California, where it had been mixed with sand and used to surface 300 m of footpath in an attempt to produce an alternative to asphalt.[375]

Solutions of calcium chloride and magnesium chloride in water had hygroscopic properties which aided moisture retention but had very short-term benefits, rarely lasting more than a season. Sodium silicate solutions created

---

[c] Also known as the World Road Association.

temporary colloidal binding. There was more success in those parts of the USA where crude oil was plentiful. Crude oil was first sprayed on unbound stone roads in Summerland (west of Los Angeles) in 1894.[376] In 1898, Los Angeles County began a program of spraying its dirt roads with crude oil. The process gained momentum in the 1920s and the so-called *oiled roads* soon came to be widely used. The Californian oils were very suitable, but the higher viscosity of crude oils from other American oilfields led to a number of failed applications. The oil was placed on the pavement by hand-led or vehicle-drawn sprays, and the product was strongly supported by the American oil industry. A later development used compressed air to ensure that the oil was sprayed at a uniform rate (see Figure 14.2). Tars were also used in a similar fashion.[377] Waste oil was sometimes tried but it was potentially toxic and damaging to the environment. More recently, it has been found that some oil industry by-products are surface-reactive ("ionic") and can act as a stabiliser (Chapter 18).

Two more practical responses were the application of tar macadam in Britain (Chapter 15) and the now widely used spray and chip seal surface to be discussed in Chapter 15. However, when writing in 1910 an expert from the American Office of Public Roads optimistically thought that the only solution to the dust problem was to treat new and old pavements with "chemical substances".[378]

Chapter 10

Using masonry methods to produce pavements for heavy traffic

All the pavements we have been discussing to this stage have had surfaces composed of individual pieces of stone or timber. A particular demand for urban street pavements was the need to be able to remove the unpleasant consequences – particularly excrement and urine – produced by animals while hauling vehicles. It was not easy to do this with a surface consisting of inherently loose stones, even when the voids between the stones were filled with smaller particles. Until the 1920s (Chapter 15), during much of the year urban pavements would be either *an ocean of mud or a desert of infected dust*.[379] Street cleaning was an ongoing task and men with straw brooms made a living as street sweepers and could be present at points where pedestrians crossed the street.[380] It was an old profession dating at least from the times of ancient Rome and is explored further in Chapter 12.

As traffic increased, there was thus a growing demand for surfaces that were impervious and could be readily washed and cleaned. On the other hand, the surfaces had to provide sufficient friction to meet the traction needs of passing traffic. In the mid-19th century, the new urban macadam surfaces discussed in the previous chapter, although a great improvement on their predecessors, also had to carry increasing volumes of traffic. They were frequently abraded, dusty in summer and slippery at all times due to the waste from passing animals and neighbouring houses.[381] Thus, a major aim of many of the alternative paving methods was to provide a good running surface which was firm in winter, produced no dust in summer, and from which animal and domestic waste could readily be removed.

Flagstones are thick flat stones capable of independently carrying traffic loads placed on their surface. Finishing the surface of the pavement with carefully placed flagstones had been practiced since Mesopotamian times and well-used by the Romans, particularly in Rome itself. Flagstone paving could be very effective but it was an expensive solution as it required special stones, masonry skills and a firm underlying pavement structure. Thus, flagstone paving was also used in Córdoba in Spain in 850 AD during the reign of Caliph Abdorrahman II, in Paris in 1184 at the instruction of Phillip II, in 1415 in the Free City of Augsburg in Germany and in 1417 a few London streets were paved with flagstones (although Chapter 7 mentions some paving prior to 1276).[382] Remnants of the Parisian pavements were found in the garden of the Musée de Cluny (near Rue du Petit Pont) and comprised flagstones with sides at least a metre long and about 200 mm thick.[383]

Commonly, the flagstones were approximately rectangular or wedge-shaped, with a slightly convex upper surface. The joints between them were usually packed with earth. Generally, it was difficult to get large flagstones to stay in position when wheels or hooves loaded them near their edges – this could cause the stones to crack or to disconcertingly squirt water upwards from between the joints. They were also prone to frost and thaw damage due to freezing water (Chapter 1). Thus, over time, large flagstones were rarely employed.[384]

The Inca civilisations in South America had extensive path systems with a total length of about 23,000 km. However, as they had no wheeled vehicles, the paths were simply paved and the greatest length of stone paving stretched for about 20 km. Their use of flat stone paving was more to prevent water scour rather than to increase traveller's comfort. Long sequences of stone steps were used in hilly areas.[385]

Smaller smoothly rounded stones known as cobblestones (Figure 10.1) or cats-heads were widely favoured as they were often easily obtained from river beds and gravel pits. However, they were difficult to place securely, were noisy under animal-drawn traffic, were often dislodged by traffic, and collected waste and mud. A 19th-century commentator described them

Figure 10.1 Worn cobblestone pavement.[389]

as *bone-shaking and ear-splitting*.[386] One advantage of cobblestone (and similar) paving was that horses, via their horse shoes or hooves, could often use the valleys between cobbles to obtain better traction.[387] Although common in many cities, they were being rapidly replaced by the end of the 19th century, due to the increase use of asphalt and brick paving and the decreased cost of broken stones.[388] In addition, they received some very poor press. The first edition of an American magazine, Paving and Municipal Engineering, published in 1890, had an anonymous article on its first page headed *The Accursed Cobble-Stone*. It read, inter alia, *The cobblestone is one step beyond barbarism. There is no harmony between civilisation and cobblestones. They represent a kind of brutal ignorance.*

A more effective solution was to shape large stones and flagstones into large blocks and smaller stones and cobble stones into smaller cubical blocks called setts or Belgian blocks (they were widely used in Belgium where the local stone was very suitable for block making). Cubical blocks about 150 mm in size were being used in Paris from at least 1415.[390] Large blocks were used in Westminster in London in about 1761. The effectiveness came at a price as a pavement using setts was about ten times the cost of a simple macadam pavement. From about 1850, smaller setts came to be widely used in Europe.[391] They were from 100 to 300 mm deep, with horizontal dimensions of between 75 and 150 mm, making them closer to the contact size of hooves and wheels.[392] Under such traffic, setts had a life of about 20 years. Granite setts were particularly favoured as they were durable, often as hard as steel and did not polish or wear under traffic. Initially they were often placed on a bed of compacted 300 mm layers of gravel, sand or chalk, but by the end of the century a similar thickness of concrete was commonly used.

Setts became infamous when used by insurgents as street barricades, as highlighted in Victor Hugo's Les Misérables which describes a three-story high barricade which was *a composite of paving stones and rubble, timbers, iron bars, etc.*[d][393] Nevertheless, by 1870 over 75% of Parisian streets were paved with setts.[394] Figure 9.5 showed setts as placed in the Strand in London in 1851. A granite sett arrangement used outside Euston Station in London consisted of three 100 mm layers of coarse gravel, gravel & chalk and fine gravel and chalk, topped with 25 mm of sand on which the granite setts were placed. The sett size was 100×75×75 (depth) mm. It performed well, was extremely quiet and its joints provided a good foothold for horses.[395] Thus, it was widely copied. In 1936, stone setts were being replaced in heavily trafficked parts of London by cast-iron blocks.[396]

Setts were conventionally arranged in a pattern that left the predominant line of joints transverse to the traffic direction. This avoided the excessive wear that occurred in longitudinally oriented joints, although some

---

[d] More recently, in 2019 protestors in Hong Kong used concrete paving blocks (Chapter 17) as missiles in their confrontation with local riot police.

Figure 10.2 Stone setts on the Champs Elysées in late 20th-century Paris.[399]

practitioners favoured a diagonal arrangement. Many urban areas arranged the setts in a fan-tail (Figure 10.2) or herring-bone pattern, although Aitken saw little technical virtue in the process.[397] Indeed, there was great debate over the arrangement of setts, much of it based on empirical observations or traveller hearsay.[398]

Flagstones, blocks, cobble stones and setts were not complete pavements and had to be placed on an underlying pavement course or a well-shaped firm subgrade. Their required depth was determined by the rate at which they wore under traffic, rather than by any load-carrying and associated structural considerations. Other than for cobble stones, their horizontal dimensions were dominated by the need to avoid any tipping when a heavy load was applied to one edge and by the need to keep the joints between each block close enough to provide continuous footholds for horse traffic. Joints filled with bitumen performed well.[400]

Many of the 19th-century pavements predominantly served as surfacings rather than as load-carrying structures and in retrospect it often seems that inadequate attention was given to these foundation courses. If the subgrade was not sufficiently firm, its strength and stiffness could be enhanced by rolling layers of broken stones into its surface in the manner of a Roman agger (Chapter 6). This approach became much more practical after the advent of heavy road-rollers. In older cities, road-makers often made these layers by using stones taken from the earlier pavement. By the latter half of the 19th century, it was more common to use an underlying foundation course comprised of 150 mm of concrete covered by a 50 mm sand-bedding layer. It was also quite common to place wooden planks over the subgrade to provide the required support, despite the fact they often subsequently rotted and warped and consequently distorted the pavement surface.[401]

In the 19th century, there was increasing used of sawn wood blocks to provide the pavement surface. This technique was developed in Russia and the Nordic countries where there were plentiful supplies of softwood from deciduous trees such as firs and pines. An advantage that the block had over the plank was that it presented the much tougher end-grain to the traffic. Consequently, wear was reduced by about 75%.

Wooden blocks were used by Gourieff (a Member of the Russian Academy) on Great Sea St and Million St in St Petersburg in Russia in 1820. He placed hexagonal blocks on a layer of broken stones and sand. The surface was then covered with tar as a preservative and sand to improve traction.[402] This was sometimes called the *Gourieff overlay*. St Petersburg hexagonal blocks were exported for use in street paving in Manchester in 1838. Wood block paving was tried in New York and Philadelphia[403] in about 1835 and in London in front of the Old Bailey in 1839 and on The Strand in 1840. Wood blocks were the best-performed of a variety of paving systems tested in Oxford St, London in 1838–1839.[404] These softwood (pine) blocks were relatively cheap but not very durable as under traffic and moist conditions a typical wood block pavement lasted from 3 to 10 years. They were also proved very slippery for horses. They thus failed to displace the longer lasting granite setts from the pavement market.

To reduce their deterioration in wet conditions, wood blocks were impregnated with tar or creosote before placing on the road surface. The method was first used in England in 1841.[405] By the 1860s, tar-treated wood block paving was widely used in Paris and Washington and, to a lesser extent, in London and Boston.[406] One consequence of the tar and creosote treatments was that the treated pavement could become quite flammable and was the cause of some major urban fires. This was the case with plank roads in San Francisco in 1850, after a Democratic Party election celebration, and with wood block paving in New York in 1870 and during the Chicago Great Fire of 1871.[407] This led to the widespread abandonment of creosoting, particularly as it was found to have minimal effect on the life of the softwood blocks.

Two advantages of wood blocks in urban areas were that they created very little noise from metal-shod hooves and wheels and that they were perceived by some to be safer for horses than all the alternatives (see Table 10.1). The blocks were typically cubic with side dimensions of 150 mm or less. Many often-commercial arrangements were used to attempt to reduce their slipperiness but usually proved ineffective and made the blocks noisy and increased the haulage task.[408] As with stone blocks, wood blocks were placed on layers of sand, broken stones, bituminous mastic (Chapter 11), mortar or concrete. Because they could be cut to accurate dimensions, they could be placed with minimal inter-block jointing; however, many builders preferred joints of about 5 mm in width, filled with cement, in order to improve the traction the joints provided for horses. The blocks could wear down under traffic by 10 mm/year, which would have led to a short life for 150 mm blocks.[409] Charles Dickens gave a telling image in 1856 when he wrote:[410]

> M Gourieff introduced [to St. Petersburg] the hexagonal wooden pavement with which, in London, we are all acquainted. This, with continuous reparation, answers pretty well, taking into consideration that equality of surface seems utterly unattainable, that the knavish contractors supply blocks so rotten as to be worthless a few days after they are put down, and that the horses are continually slipping and frequently falling on the perilous highway. It is unpleasant, also, to be semi-asphyxiated each time you take your walks abroad, by the fumes of the infernal pitch-cauldrons, round which the moujik workmen gather, like witches.

A key public concern highlighted by Dickens was that many wood block pavements produced an unpleasant stench due to the use of tar and the accumulation of rotting wood and the excrement of horses and oxen. It was also publicly believed that the vapours rising from damp wood block paving were a carrier of fatal diseases such as typhoid, dysentery and malaria.[411,412] In 1877, these factors led to New York removing all of its 30 km of wood block pavements, many of which had rotted after less than 5 years.[413] A similar sequence of events occurred in Washington where the wood blocks lasted less than 4 years. The resulting scandal is said to have been primary cause for the city losing home rule in 1874 and becoming a federal District (see also Chapter 15).[414] Most of the other pavement types then in use had similar unpleasant side-effects and Table 10.1 quantifies some of the issues that were being encountered.

Despite these experiences, overall the use of wood block paving continued to increase. A variety of proprietary wood block paving methods came on the market in the last quarter of the 19th century[420] and a

Table 10.1 Hygiene problems associated with various road surfacings.[415–418] See also Table 13.2

Pavement type	Relative number of loads of mud produced	Filth rating	Horse accident rate (distance travelled in km between accidents)
macadam	1.0[a]	very poor	not quantified
setts	0.7	Poor	320
wooden blocks	0.2	very poor (softwood) good (hardwood)	550
asphalt	0.1	very good	220

[a] A typical surface produced about 20 mm of mud per week.[419]

few are illustrated in Figure 10.3. In 1880, Adrien Mountain conducted a trial of eight different types of wood blocks in Sydney. The hardwood blocks in the trial performed very well.[421] As a consequence of Mountain's work, softwood blocks in European cities were often replaced by hardwood (eucalyptus) blocks imported from Australia.[422] Their first application was in 1888 when they were used to pave the road on London's Westminster Bridge over the Thames. The hardwood blocks were very durable and could last up to 12 years but had a propensity to shrink and therefore leave jointing gaps where chipping under traffic resulted in a noisy surface. American pavement-makers found local cedar and cypress blocks to be almost as durable as hardwood.

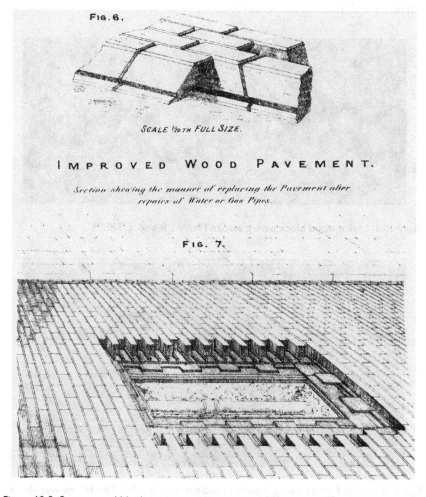

Figure 10.3 Some wood block paving variations in an 1870s review.[425]

With the widespread use of hardwood-type blocks and small (c5 mm wide) joints, by the end of the 19th century wood block paving was, internationally, the preferred method of urban paving and 15 year lives were being achieved.[423] They held this preeminent status until the 1920s. An example of block placement from Rome in about 1910 is shown in Figure 10.4. In 1913, Mountain presented data at an international conference showing that pavement lives of 30 years could be achieved, if some maintenance was performed after 20 years.[424] As concrete placement technology improved, it was possible to form the surface of the concrete supporting course such that the wood blocks could be placed directly on the concrete, thus avoiding problems with the sand cushion layers moving or permitting water movements. It was also found valuable to fill the joints between the blocks with bitumen or tar to further prevent water intrusion. One major effect of water intrusion is illustrated in Figure 10.5.

Figure 10.4 Laying wood block paving on San Pietrini, Rome, c1910.[426]

Figure 10.5 Wood block paving failure in Indianapolis in 1917. The failure was due to hydraulic pressure in water between the blocks and the basecourse.[427]

Apart from durability issues, a growing problem with wood block paving was maintaining a surface that over time had enough surface friction to meet the braking and turning needs of the new faster-moving internal-combustion vehicles, as suggested in the rankings in Tables 10.1 and 13.2. The smoothness problem was partly addressed by spraying the surface with bitumen sprinkled with grit or sharp sand. Minneapolis, the home of many timber industry magnates, made extensive use of creosote-soaked wood blocks between 1902 and 1912. After 1920, in most areas wood blocks gradually lost their premier status to asphalt and concrete, although they were the major form of functioning urban paving in many cities prior to World War II. Paris constructed its last length of wood block paving in 1938.[428]

Depending on local costs, it was sometimes more realistic to use burnt clay bricks, brick-like paving blocks or concrete bricks. Brick paving dates from the earliest Mesopotamian pavements (Chapter 2) with more recent use beginning in The Netherlands in the 18th century and Edinburgh in 1866.[429] The Mesopotamian usage resulted from the absence of suitable rock.

From the 19th century, bricks have been widely used for pavements in towns, including many the Low Countries of Europe and in the Mississippi valley in the USA. The Low Countries, and particularly the Dutch use, resulted from much of these areas having at least 10 m of compressible soil. Large pavement settlements were inevitable and block paving could be quickly and cheaply relevelled. The American use occurred where clay was much easier to obtain than broken stone.[430] The associated pavement design methods have largely been empirically based on past service experience. Bricks and concrete blocks still find some predominantly aesthetic use as a surfacing layer in modern urban street pavements. Modern bricks are usually pressed and oven fired and sometimes glazed (vitrified) to make them less pervious to water.

One initial problem with brick paving was that some of the cheaper bricks expanded with temperature rises, and this would cause the whole layer of brick paving to buckle upwards or push kerbs out of their horizontal position. A similar problem occurred in cold weather when water trapped below the bricks froze and expanded. Even a slight upward buckle of the brick surface could cause a noticeable drumming noise as vehicles travelled along the pavement. Bricks are now typically closely packed horizontally but where joint filling is required, sand, cement or bitumen have been used.

The demand for bricks created by building reconstruction in European cities following World War II led to the use of concrete blocks for paving, particularly in The Netherlands in the 1950s. A major advance came with the use of blocks somewhat intricately shaped to allow them to interlock together horizontally. It was much easier to produce such shapes with concrete than with the clay-based mixes used for bricks. Machinery for making and placing interlocked concrete block paving became available in the

1980s, and the initial major applications were in West Germany and The Netherlands. Concrete block paving is now widely used and has a number of advantages over other paving in that it:[431]

- provides easy access to underground services,
- can be placed by unskilled labour, and
- can be provided in a range of colours.

However, it is noisier than conventional paving, particularly at higher traffic speeds.

Contemporary paving machines can lay brick and block paving up to 5 m wide in a single pass of the machine (Figure 10.6). One advantage of concrete blocks is that, if no longer needed, they can be readily recycled and used in new pavements.

It is necessary to fully waterproof a block pavement to prevent water from damaging the underlying course or causing uplift pressures (Chapter 1). One technique based on earlier block paving was to place fine sand or mortar in the joints between the blocks. However, movement of the blocks under load can readily displace such a jointing material. Modern practice suggests that 'shaped' blocks made with tight tolerances and placed on sound

Figure 10.6 A concrete block-laying machine in operation at the European Container Terminal in Rotterdam in the mid-1980s.[433]

basecourses and with only small gaps (3 mm) between the joints result in the joints quickly becoming sealed naturally with detritus which prevents water entry.

Blocks made by compressing powdered rock asphalt were first used in Paris in 1824 and are discussed in Chapter 13. Other methods of block paving included blocks made of glass at one extreme and compressed sewage at the other.[432] It should be reemphasised that these block-type pavements were not self-supporting and needed to be placed on a structurally competent lower pavement layer. Rammed sand was often used to ease placement and the sand was placed on concrete if the subgrade was inadequate. The other advantage of using sand rather than concrete as a basecourse with block paving is that it made it much easier to place or repair services beneath the pavement surface.

As iron-making capabilities improved in the 19th century, there were a number of attempts to use iron to produce useful pavements. The most common was the plateway in which two running strips were placed to carry the wheels of carts and wagons.[434] An early predecessor was a timber plank road built in 1758 to carry coal from Middleton to Leeds in England. A more extensive timber system in North America is described in Chapter 5. Sometimes granite blocks were similarly used to aid animal-drawn vehicle on steep hills. A good plateway could increase the haulage capacity of such vehicles by an order of magnitude. Thus, when large-scale iron-making began at Coalbrookdale in England in 1767, an early plateway application was at the iron works. The usage then expanded to other ironworks, aided by the use of flanged plates to ensure that the wheels did not leave the tracks. A major extension occurred with the Surrey Iron Railway which operated between Wandsworth and Croydon in Surrey between 1803 and 1846. Users supplied their own vehicles and paid a toll to use the plateway. The system was relatively expensive to operate and maintain and in England could not compete with the railways. Nevertheless, it was spasmodically used in many locations around the world, including New York, but was swept away by the introduction of high-speed motor vehicles early in the 20th century.

On the Holyhead Road in the 1820s, Telford's assistant, John Macneill, developed and patented embedding 25 mm cubes of iron in the pavement surface at 100 mm centres. It was not clear what Macneill had in mind and apparently there was no positive outcome.

Prior to 1855, there had been a trial of iron paving in Glasgow. The iron blocks and the alternative granite blocks were about the same cost.[435] Other trials of iron pieces that were smaller than a horse's hoof were unsuccessful as the horses often stumbled and fell. Similar problems occurred when large iron plates were used for surfacing. They were briefly reintroduced in England in the 1930s.[436] Iron block paving was used on the Unter den Linden in Berlin between 1877 and 1890.[437] A mesh-like iron layer (Figure 10.7) was successfully used in Paris in the 1930s.[438]

Figure 10.7 Iron paving in Paris in the 1930s.[442]

An opposite approach tried in Illinois in 1930 was to place steel sheets directly onto the subgrade. The sheets were then covered with a layer of bituminous mastic (Chapter 11) which in turn supported a surface layer of bricks.[439] In retrospect, it is difficult to see the merits of the approach.

An alternative using interlocking, gravel-filled, circular iron cells 300 mm in diameter and 150 mm deep was developed by General Krapp in America. In the 1850s, the method was used on Nassau St in New York, in Boston and on Leadenhall St in London.[440] The idea resurfaced in Darien, Connecticut, in 1943 where some streets were paved with steel grating panels measuring about 4 m by 600 mm placed on a subgrade. After being placed, the panels were filled with sand and then coated with oil.[441]

Many other materials have been trialled as potential pavements. Rubber slabs were tested on the Admiralty Courtyard in London in 1840. They were then used on concrete basecourses outside London's Euston and St Pancras railway stations from 1870 until at least 1934, in 1913 in Southwark, London and at Leeds railway station until the 1950s. The rubber slabs required regular replacement.[443] Their purpose was to reduce traffic noise, although this need would have diminished as pneumatic tyres became the norm. Nevertheless, rubber block paving was used in the Mersey Tunnel in Liverpool in the 1930s. Leather paving was also tried as a noise-reducing measure.

An extensive review of 19th-century segmental pavements is available in later editions of Law and Clark's 19th-century textbook on roads and pavements.[444]

Chapter 11

The essential paving properties of bitumen, asphalt and cement

The definitions and terminology that this book employs in discussing soil, sand and stone were visited in the Detour in Chapter 4. This Detour visits similar aspects of bitumen, cement and asphalt, particularly as there are many terminological variants and misconceptions with respect to these materials in common use. It can be skipped by readers who already know, or who don't wish to know, about the technology of these pavement materials.

BITUMEN

The history of bitumen was discussed in Chapter 2. Although bitumen will figure frequently in this story, its characteristics are not widely understood.[445] It is sometimes called *asphaltic cement* or *bituminous cement* in North America.

While bitumen can occur naturally, today it is most commonly produced as a residue during the distillation of natural crude petroleum where it is the highest boiling point fraction.[446] Bitumen is composed of hydrogen and carbon and is thus a *hydrocarbon*. It is a mixture of many chemical compounds broadly known as association colloids and which are loosely held together by hydrogen bonding.[447] By weight, it is 85% carbon and it may also contain small amounts of sulphur. Its actual structure will depend on its source oilfield and can even vary within a particular oilfield. Bitumen is not internally uniform, and about 55% of a bitumen is a hydrocarbon oil called *maltene* which has a chain-like molecular structure that gives bitumen its viscoelastic properties. The second major component is about 20% of bitumen and is called *asphaltene*. It exists in a colloidal suspension in the maltene which means that technically bitumen is a *sol*. Asphaltene gives bitumen its adhesive properties and controls its viscosity. Bitumen is lighter than water.

Bitumen is a thermoplastic adhesive which, when heated to about 150°C, readily attaches to pieces of stone. Instead of heating, it can also be made fluid by foaming with steam (Chapter 18) or by "fluxing" or "cutting" it by the addition of products like kerosene. Effectively bitumen is chemically inert, impermeable, and water resistant and is malleable and ductile over a practical range of temperatures. However, over time it can lose many of its useful properties. Most of the durability loss is due to thermal oxidation which increases exponentially with temperature. In particular, ultraviolet light causes the maltene to photo-oxidise. As bitumen is a good light-absorber, this can only occur within 5 μm of the surface of the bitumen.[448] The bitumen found in rock asphalt has usually incurred some oxidation.

The main durability issue with bitumen occurs when it is used as thin spray covering a surface layer of stone (Chapter 14). Such layers are directly exposed to the sun, and the loss of durability will directly cause a loss of stones from the surface layer. In California, in the 1960s oven-based tests of rolling films of bitumen were devised to give some guidance on its propensity to harden.[449] Subsequently, the provision and maintenance of long lengths of low-cost sealed roads in Australia led to the Australian Road Research Board (ARRB) durability test which detects bitumens which are most prone to oxidation in high ambient temperatures.[450] The ARRB test is based on the earlier Rolling Thin Film Oven tests.

This book often speaks colloquially of the *fluidity* of substances, but the technical term is *viscosity* which measures how difficult it is to make a fluid flow, it is thus the opposite of fluidity. It can be thought of as internal friction within a fluid and results from cohesion between its molecules. Bitumen is quite hard and viscous, but – as mentioned above – this can be reduced by heating it, by adding oil, or by mixing it with other materials to form an emulsion. (An emulsion occurs when two incompatible liquids are mixed together so that one liquid exists as small drops within the other.)

In each of these cases, the bitumen remains sticky. Bitumens modified in this manner are now widely used as a surface coat for rehabilitating pavements that have exceeded their serviceable life. They also have increasing use in allowing stones and bitumen to be mixed and placed at lower temperature (e.g. 140°C rather than 170°C). Asphalts using these bitumens are called *warm mix asphalts*, as opposed to the common *hot mix asphalts*, and can often result in lower paving costs and the consumption of less energy.[451]

An early way of checking the viscosity of a bitumen sample was to have a person chew a piece and compare its chewability with a sample of bitumens of known viscosity kept in "calibration box". Indeed, the Aztecs did use bitumen as a chewing gum.[452] In discussing the use of tar in paving, Deacon in 1879 had noted[453] that *foremen accustomed to the work can judge (whether it is too brittle) by biting it*. In 1888, Bowen of Barber Asphalt (Chapter 15) replaced these oral tests with a test in which a needle was poked into the bitumen.

Needle *penetration* tests of this kind were (and still are) common in testing pavement subgrades, cooking and in the checking of meat products. Bowen used a "No. 2 sewing needle", and his penetration test was standardised.[454] Modest variations of this test are also still being used today to assess bitumen, particularly its loss of ductility over time.[455] In a typical current test, the penetration of a 1 mm square needle at a controlled temperature (15°C or 25°C) and under a specified load is measured in millimetres. If the bitumen is hot and fluid, the viscosity is measured by the time it takes to pour the bitumen between standard containers. An early test of cold bitumen was to check for cracking when a cylindrical sample was hit with the ball-end of a hammer.[456]

Mastic: a mixture of bitumen and sand or of powdered asphaltic rock and sand, softened by heat or by the addition of more volatile oils or bitumens. The term came from *mastics* that had been derived from tree saps. The material certainly has some similarities with the tree-sap mastics, which also soften when heated or moulded. Mastic could be delivered to paving sites as blocks or cheeses that would then be reheated to make them fluid enough to form into a pavement (Figure 11.1).[457] The term *bituminous mastic* highlighted its use of bitumen. *Asphaltic mastic* was commonly made from powdered rock asphalt.

Petroleum: a natural material formed from the decay of dead organisms and found below the earth's surface.

Petrol: a "light" combustible material obtained by refining petroleum. It is sometimes called *gas* in North America.

Figure 11.1 Mastic cheeses. The cauldron for heating the cheeses and the tools for carrying, placing and smoothing the mastic are in the foreground.[458]

Tar:[459] a bitumen-like material obtained by burning timber or coal. *Pitch* is produced by distilling tar. The method for doing this was introduced by Friedrich Accum, a German chemist working in London in 1815.[460] Tar soon became widely available and was relatively cheap in the 19th century as a by-product of the production of coal gas for lighting, particularly street lighting, and for early internal combustion engines.[461] It was widely used for the next hundred years before its carcinogenic properties became more evident. A standard textbook[462] on tar published in 1938 somewhat blissfully commented *All tar warts are not necessarily malignant, but are characterised by relatively rapid growth.... If treatment is applied in the early stages a complete cure is certain.* Bitumen was a non-carcinogenic alternative to tar as both have otherwise similar properties and chemistry.

Asphalt: a mixture of bitumen, stone and (possibly) sand. As discussed in Chapter 15, some popular mixes are called *asphaltic concrete*, or *bituminous concrete* which implies (wrongly) that conventional concrete should be called cementitious concrete.[463] Asphalts vary from those with very small proportion of stone and sand to those where the stones are all in effective contact and the bitumen only coats the stones and fills voids in the interstices of the stone layer. These latter asphalts are the ones that came to be used as structural courses in pavements as any vertical loads on the asphalt layer are transferred through to the base of the layer by stone-to-stone contact, thus making them strong and stiff. On the other hand, bitumen-rich asphalts defined below rely on the bitumen for load transfer and are thus weaker, softer and temperature-sensitive. The word *asphalt* is of Greek origin and relates to the fact that it is not as slippery as mastics.

Hot mix asphalt (HMA): an asphalt where all the components (particularly the bitumen) have been heated prior to mixing to allow proper mixing and placement to occur. Most contemporary asphalts would be HMA and, in the USA, the term has generally replaced *asphaltic concrete* as the common name for asphalt in pavements. However, there are practical advantages in keeping the temperatures as low as possible and this has led to the development of *warm mix asphalt* where the bitumen viscosity has been temporarily lowered, commonly by using steam to produce a foamed bitumen (Chapter 18).

Rock asphalt: asphalt quarried from rock geologically impregnated with bitumen. It is sometimes called *natural asphalt*. The common paving version of the material used the rock asphalt after it was crushed to a fine powder and was called *powdered rock asphalt*. Powdered rock asphalt with added bitumen was commonly called *asphaltic mastic*. *Gritted asphaltic mastic* also contained added sand.

Cement: is made by burning limestone or dolomite to change their main constituent – calcium carbonate – to calcium oxide. Clays are sometimes added to the process.[464] Primitive cements date from about 6000 BC. As noted in Chapter 2, cement had been used by the Greeks and Etruscans in about 400 BC and then by the Romans. Modern Portland cement is made by heating limestone to about 850°C in a controlled manufacturing process which includes the addition of gypsum.[465] Due to the surface chemistry of its particles, cement is very chemically reactive when mixed with water and it then sets into a material that is strong in compression but weak in tension (i.e. it is brittle). It can be attacked by acids.

Mortar: mixing cement, water and sand produces *mortar*. The commonest mortars were and still are cementitious or bituminous (particularly if tar is considered bituminous[466]). Chapter 10 discussed using mortars to waterproof block pavements and to secure the blocks in position.

Concrete: adding stones to a mortar produces *concrete*. The Roman use of concrete is discussed in Chapter 6, and modern concrete pavements are discussed in Chapter 17. The use of concrete to produce a new pavement type was a major step which is pursued in Chapters 13 and 17.

Pavements: now fall into two categories. To this stage, we have been discussing unbound pavements. A *bound* pavement is one where all the components of the pavement act together as a single coherent material. Until they crack, concrete and asphalt are bound materials. Temperature extremes can also cause asphalt to become partially unbound.

Chapter 12

Three unique pavement engineers – Metcalf, McAdam and Mountain

To balance the various fascinating but potentially turgid technological stories presented throughout the other chapters of this book, this chapter provides some passing evidence of the human side of pavements by respectfully exploring some non-technical aspects of three of the major contributors – John Metcalf, John McAdam and Adrien Mountain – to pavement technology, before sweeping all aside in its final paragraphs.

Beginning with John Metcalf, as a disclaimer, it must first be said that this John Metcalf bears no known links genealogically or technically to his namesake who is a co-author of this book. Any common oddities and eccentricities can be fully explained by the fact that both were Yorkshiremen.

The technical contributions of the first John Metcalf are discussed in Chapters 5 and 8. Metcalf was born in 1717 in Knaresborough in the West Riding of Yorkshire to working class parents, although his father did keep horses.[467] He was blinded by smallpox when he was 6 years of age and received no further formal education other than music lessons. Nevertheless, he became well known as a very clever and strong young man who climbed trees, swam and dived in the River Nidd where he saved three people from drowning, rode horses and played the fiddle at public events. He was active in horse trading, horse racing, fox hunting, cock fighting, bowling, smuggling and gambling. In horse racing, he kept on course by listening for dinner bells rung by friends situated around the course.

After being correctly accused of fathering a child by a local unmarried girl, he took himself to Whitby and hence by ship to London. There he met Colonel Liddle, a member of parliament for Berwick, who offered Metcalf a ride in his coach back to Knaresborough, where he usually stayed for three weeks using the town's famous spas. Metcalf declined and said he would walk back. Blind and having never made the 320 km trip before, he arrived home 2 days before the Colonel and his coach and 15 servants. Metcalf subsequently married the daughter of the wealthy landlord of the local Royal Oak hotel, having successfully proposed to her and then eloped together on horseback the night before her intended marriage to another man. They married the next day in a nearby town. Her parents were very angry when they learnt of the marriage. Metcalf first met his mother-in-law formally when, to win a bet, he rode his horse into the bar of the Royal Oak hotel. The marriage was fruitful and at the time his death in 1810 at the age of 93, Metcalf had 90 great-grandchildren (Figure 12.1).

There are many other stories about Metcalf.[469] One that illustrates another aspect of his character occurred in 1745 when a local Squire enlisted him as a recruiting sergeant to gather volunteers to join the Royal Army to fight in the north against Bonny Prince Charley's Scottish rebellion. Metcalf led his 64 volunteers playing on his fiddle. They joined the English army in time to be part of its defeat at Falkirk. In the process he incidentally met General Wade (Chapter 3). Metcalf escaped by pretending to be a fiddler engaged to entertain the Prince. Many English officers also "escaped". When an officer from another regiment formally asked Metcalf how a blind soldier found escape so easy, he replied *I found it very easy by following the sound of the Officers' horses (as they retreated from the Highland men).*

In 1751, he began operating a wagon service between Knaresborough and York and this led to his interest in roads, beginning with a turnpike between Harrowgate and Boroughbridge. His work on roads was discussed in Chapter 9.

An even more prominent contributor to pavements was John McAdam (Figure 9.1). As noted in Chapter 9, McAdam's name has passed into the literature as *macadam*, the form of pavement he invented, and a later

Figure 12.1 John Metcalf, aged 78. [468]

form called *tar macadam* soon became *tarmac*, a word still widely used to describe the pavements at airfields. But on two grounds the McAdam name was probably a fiction. First, he wrote his name M'Adam. Second, stories have it that the family name had been McGregor, but life-threatening troubles with their English overlords meant that the family had to flee its home and a new name had to be found. As strict Scottish Calvinists, their consciences were assuaged by the belief that they did not tell a falsehood if they informed any enquirer that their name was McAdam, which literally means "son of Adam".[470]

McAdam was an innovative but difficult man who made many very public and very acerbic comments about the abilities of his contemporaries, for instance:[471]

Road engineers
- deceive the public and delude the trustees;
- elected because they can measure; they might as well be elected because they can sing;
- their defect lies in the want of science and has gone hand-in-hand with improvident expenditure; without energy or character.

Trustees of private roads
- narrow and jealous.
- their accounts: obscure, confused, perplexed and dilapidated.
- their road-making: a great waste of public money.

Public servants
- worthless.

Wood block paving was discussed in the previous chapter where it was seen that the product encountered many pitfalls. A major step forward occurred when Adrien Mountain introduced hardwood blocks which he had successfully used in his Australian homeland (Chapter 10). It was a fairly obvious local innovation as most Australian trees are hardwood – whereas softwoods predominate in the northern hemisphere. One of Mountain's key attributes was the enthusiasm with which he pursued the use of hardwood blocks, first in Australia and then internationally. Mountain's work was summarised technically in Chapter 10. Here, we will explore some less technical aspects of his endeavours.

Mountain was educated in Sydney and became its City Engineer before moving to Melbourne as its City Engineer in 1886. He became President of the engineering professional body in Victoria and was awarded a Gold medal for one of his papers. He also published a book on wood block paving.[472] Mountain had a well-deserved reputation for hard work, intellect and a love for his profession.[473] He was not without some personal vanity and a Melbourne satirical magazine in 1912 described him as *a diminutive man with a large brain, ... and a plaintive voice.* He apparently liked to highlight his resemblance to King George V and awaited the new copies of English magazines depicting the King. He then dressed himself in a similar fashion[474] (Figure 12.2).

When Mountain successfully introduced wood block paving into Sydney, it caused consternation among powerful Sydney importers of asphalt, who saw their product pushed aside and their profits savaged. Mountain records[476] that they had led *a fierce and powerful attack* on wood block paving. They lobbied NSW parliamentarians on the unhealthy nature of wood block paving and were rewarded by the Government in 1884 establishing a Wood Paving Board to inquire into the *alleged deleterious effects of wood paving on public health.*

The Board called two key witnesses, a local asphalt supplier and a local doctor who had previously practised in India, where he ran a hospital. He apparently testified that initially many of his Indian patients had died in his hospital due to diseases such as malaria and yellow fever. He had noticed that his staff were regularly washing the hospital's wooden floors with water and deduced that the vapours from the wet floors were the vectors that spread fatal diseases to his patients. He ordered that the staff stop washing the floors. Very soon patients stopped dying from malaria and yellow fever. This mightily impressed the Board who noted as an aside that

Figure 12.2 Adrien Mountain in Sydney (1881).[475]

the floors of hospital wards should never be wetted before returning to its given task and recommended banning the future use of wood bock paving in NSW. In doing so, the Board ignored the likelihood that most of the patients in the now unwashed Indian hospital would by then probably have been preferentially dying from golden staph. It also ignored the submission by Mountain – the only witness called to support block paving – that wood block paving was widely used in Louis and Marie Pasteur's own city of Paris, without any concerns having been expressed by those two eminent microbiologists.[477]

In fairness to the doctor, it must be said that his erroneous view that malaria was *the product of heat, moisture and vegetable decomposition* was commonly held at the time. In addition, he was not alone in his concern. Despite the evocation of the Pasteurs, a French professor of naval hygiene, Jean Baptise Fonssagrives, believed that noxious miasmas would arise from a pavement's wet wood surface and that a city paved entirely with wood would become a *city of marsh fevers.*[478] In addition, it was a widespread belief that washing the wooden floors of hospital wards did spread gangrene.[479]

The Board's findings were largely ignored but Mountain nevertheless, in high dudgeon, moved to Melbourne in 1886. By 1907, some 50 million wood blocks had been used for street paving in Sydney and Melbourne. Mountain might also have been helped by some advice he received in 1880 from E. J. Harrison, then Secretary and Manager of the Val de Travers Company in London. Harrison advised Mountain, inter alia, that *Frenchmen and Italians make the best compressed layers.*[480]

As will be discussed in Chapter 13, wood block paving was being replaced by asphalt paving in the 1920s, possibly inspired by the famous British innovator, Heath Robinson, although his 1923 solution shown in Figure 12.3 curiously never came to fruition. Robinson was acquainted with some leading pavement engineers.[481]

The matter of the cleanliness of pavements was a very old problem. Street cleaning has been an ongoing task dating at least from the times of ancient Rome. It was thus an old profession and in most cities men with straw brooms made a living as street sweepers at points where pedestrians crossed the street.[482] In England, they were called by some variant of the title "scavenger" (Figure 12.4). Henry Mayhew in his famous catalogue of 19th-century English workers wrote *Crossing sweeping seems to be one of those occupations which are resorted to as an excuse for begging.*[483] Another of his characters, Billy, commented that he *couldn't abide this mucky-dam* (macadam) *... its sloppy stuff and goes bad in holes.*

In 1184, the stench of Parisian streets so offended Phillip II when standing at the window of his palace that he ordered all of its streets to be paved so that it would be possible to remove the daily accumulation of excremental muck that was deposited by transport animals and by adjoining citizens who dumped all their personal and domestic waste and dead domestic

Figure 12.3 Heath Robinson's 1923 asphalt paving machine.[484]

Figure 12.4 London street scavenger, c1850.[487]

animals into the streets. From 1348, abutting owners were required to pay for Parisian street cleaning, in 1389, this was extended to the nobility and, in 1399, to members of religious orders. From 1518, the cleaning task was undertaken by petty criminals bound in iron chains.[485]

It may be appropriate to note some national differences on the matter of dumping household waste and excrement into the street. Laws were introduced to manage the way the waste was tipped into the street from upstairs windows. An ordinance in 1502 prohibited Parisians from tipping waste out of windows but this was gradually relaxed to a requirement that they shout *gardez l'eau* before tossing. However, in Scotland, the obligation was placed on the pedestrian who had to shout *don't throw* in the local brogue if he thought it appropriate.[486] Flower boxes on window sills were popular ways to conceal any inappropriate tossing.

The problem also existed in the New World. *General Order No. 28* was a military decree made by Major General Benjamin Butler during the American Civil War. Following the Battle of New Orleans, Butler established himself as military commander of that city on 1 May 1862. Many of the city's inhabitants were strongly hostile to the Federal government, and many women in particular expressed this contempt by insulting Union troops, including throwing the contents of their chamber pots over them. Some also printed his portrait on the bottom of their chamber pots, and this was considered to be a cause of Butler's removal from command of New Orleans on 16 December 1862.[488]

The next chapters explore the long and extensive role of bitumen in pavements. In Chapter 11, we have already noted that its viscosity was checked by chewing a lump of bitumen and comparing its chewability against a box of calibrated pieces of bitumen. A mixture of bitumen and chicle has long been a popular chewing gum in Mexico. Somewhat similarly, human stomach ulcers were treated by drinking an emulsion of beer and bitumen. At another extreme, bitumen was also used to enhance the taste of wine. Perhaps best of all, according to the Apocrypha, Daniel defeated a marauding dragon by feeding it with an incendiary mixture of bitumen and hay, which was ignited by the legendary fire in the dragon's belly, thus causing a severe attack of heartburn which destroyed the dragon.[489]

Undoubtedly, the most expensive paving material will be the transparent gold that Chapter 21 of the Book of Revelations tells Christians will be used to pave the street of the holy city, the new Jerusalem, when it comes down out of heaven. And it will be no mundane gold, but *pure gold, as transparent as glass*.[490] There may have been some trial runs of gold paving, for George Colman Jr in his somewhat bawdy 1797 play, Heir at Law, includes the verse:

Oh London is a fine town
A very famous city,
Where all the streets are paved with gold,
And all the maidens pretty.

Chapter 13

The first asphalt pavements, produced in 19th-century France and England

As we saw in Chapter 2, both cement and bitumen have been used as pavement surfacings and sealants since Mesopotamian times. Pavement making was relatively straight-forward in oil-rich Mesopotamia where petroleum, bitumen, sand, stone chippings, man-power and money were all readily available. There is archaeological evidence of paving mastics being produced by mixing hot bitumen and sand. In about 1200 AD Yaqut al-Hamawi, a well-known Arab biographer and geographer, wrote of the bitumen pit at Dir al-Qayyara near Mosul:[491]

> There are workers who collect bitumen from the spring in woven reed baskets and pour it over the ground. They also have large iron kettles placed over cauldrons which they load with known proportions of bitumen, water and sand. When the stirred mixture reaches the right consistency, it is poured over the ground as pavement.

Yaqut al-Hamawi's text implies that this method of paving roads was used throughout the region in medieval times.

However, technically Yaqut al-Hamawi is describing a mastic rather than an asphalt and there is no evidence from these earlier times of the use of asphalt, which requires adding stones to the mastic. Nevertheless, it seems improbable that the process he describes was not extended by adding stone chippings to the hot mix and so the first use of asphalt paving probably did occur in Mesopotamia prior to 1200 AD. Nevertheless, there is no record of asphalt paving being used anywhere in the world before the 18th century. Unlike stone paving, bitumen is not a simple material to use as it must be separated from host materials. This requires a lot of heat, which is also needed to keep the bitumen fluid for mixing with stone and sand and for placing on the pavement. The first recorded European use of bituminous paving was in 1747, when a mixture of ship's bitumen (probably from Trinidad Lake, see below), gravel and rubble had been used to produce an asphalt carriageway in the Dublin home of the Earl of Leinster. The method is believed to have been copied from a carriageway in the estate of a Prussian nobleman.[492] There were a number of petroleum sources in Prussia.

Coal tars were often a valueless waste product. Tar mixed with earth and fibres was used for paving in 1832 in Gloucestershire, England.[493] It performed well for a number of years: John Rennie – the well-known bridge designer – was the responsible engineer.[494] Tar had been used for making a footpath on the promenade at Margate Pier in Kent in England in 1822. The hot tar was poured over a macadam pavement, filling its interstices and then covering the stone layer to a depth of 5 mm. The hot tar surface was then sprinkled with sand to avoid a slippery finish.[495] One European alternative application of this process was to coat the broken stones with tar before placing them on the subgrade, and to then roll the layer of stone and tar to produce a macadam pavement. Alternatively, tar was added to an in situ layer of stones, bringing the tar to a level about 10 mm below the top of the stones. The layer was then rolled until the tar was at the top of the stones. This waterproofed the pavement and helped hold the stones in their compacted position. Later, bitumen was sometimes substituted for tar.

In the 1830s and 1840s, Antoine-Rémy Polonceau made many contributions to French pavement practice.[496] In 1834, he placed about 30 mm of tar on the surface of a macadam pavement in Paris. The tar typically penetrated up to 50 mm into the interstices in the stone layer. The first tar penetration macadam pavement in England was placed on a 3 km length of Lincoln Road outside of Nottingham in 1840.[497] It performed poorly in service, as did other similar trials. In cold weather it was necessary to heat the stones as well as the tar in order to keep the tar fluid enough to penetrate into the stone layer. A version of this mix was developed in Australia where there was no locally available bitumen or asphaltic rock. However, tar was available from the burning of coal to produce gas and the carbonisation of coal to produce coke for iron-making.[498] Tar-based trials on Vermont Ave in Washington in the 1870s were unsuccessful,[499] although Fred Warren blames the inexperienced contractors for the failure.[500] A key problem with the penetration method was that it was difficult to determine how far the bitumen had penetrated and so some poor pavements were unknowingly produced.

In practice, and in the absence of large pieces of construction machinery, these penetration methods were far from simple processes and a major authoritative text in the early 20th century devoted a 60-page chapter to the various predominantly manual methods that had been used in these penetration processes, with varying degrees of success.[501] Tar penetration macadam gained some technical and linguistic fame in 1901 when Purnell Hooley observed that a barrel of tar which accidentally burst over a layer of blast furnace slag on a road near the Denby Iron Works in Derbyshire produced an admirable road pavement. After some further investigation, including adding pitch, cement and resin to the tar, Hooley produced a useful pavement and patented his product in 1902.[502] In 1903, he laid a length of the product, now called Tarmac, near Trent Bridge in Nottingham and founded the Tarmac Company to market the new product, based largely on its specific use of tar and blast furnace slag.[503] The Company still exists and the word *tarmac* is now used as an alternative to *pavement*, particularly for airport pavements. Tarmac-type pavements using coal tar were successfully constructed in Rhode Island, USA, in 1906.[504] The invention process patented by Hooley had actually independently occurred at a number of separate locations.[505] By 1930, Tarmac were acknowledging that the market was moving from tarmac to the hot rolled asphalts to be discussed below.[506]

The resulting pavements were a form of *asphalt*. Over time, *asphalt* has had a variety of meanings although today it would commonly be taken to imply a manufactured mixture of bitumen, broken stone and sand.[507] Originally, in Europe *asphalt* referred to naturally occurring unconsolidated limestone rocks impregnated with bitumen – a material which is now commonly called *rock asphalt*.[508] Many of these rocks would readily soften with heat and crumble with pressure. Geologically consolidated rocks impregnated with bitumen but which were much harder to soften and crumble were called *bituminous rock*. Limestone and petroleum are both formed from the remains of dead creatures, so their co-occurrence may not be coincidental. However, sandstones impregnated with bitumen have also been found in many countries. Indeed, bituminous deposits occur throughout the world, ranging from almost pure bitumen, as in the Mesopotamian pits referred to earlier, to rocks containing only a small proportion of bitumen.

The use of these natural asphalts for paving had an accidental birth. In medieval Europe asphalt had long been quarried and then heated to obtain bitumen, which was primarily used for waterproofing and for animal and human medicinal uses ranging from covering major wounds and amputation sites to treating gout and rheumatism.[509] For example, sheep shearers douched the wounds they inflicted on sheep with a covering of hot tar, as told in the Australian folk song Click go the Shears and its call of *tar here Jack* to the tar boy. Natural bitumen, which oozed from a spring at Pitchford, about 10 km west of Ironbridge in England, was marketed as Betton's British Oil to cure rheumatism, skin sores and ulcers.[510]

Pavement serendipity occurred when asphalt pieces fell from a quarry truck and were compressed into the stony quarry access path by the iron-rimmed wheels of the truck. The quarry operators found that this accidental process left behind an excellent piece of paving. At least 13 such accidental discoveries of asphalt paving have been reported and so some may be apocryphal.[511]

Rock asphalt deposits occurred throughout Europe.[512] The rock was heated to make the bitumen it contained fluid and able to be separated from the host rock, although for some deposits the rock had first to be crushed – often to a fine powder – before the bitumen could be separated. The bitumen was then sold for a range of purposes, including as a medicine[513] and as a waterproofing material.[514] Of particular relevance to this story is the rock asphalt mined in the Jura Mountains at Seyssel in France

Figure 13.1 Cross-section of Seyssel asphalt mine.[516] The entire cross section is limestone but the seams impregnated with bitumen are 2–3 m thick.

(Figure 13.1) and at Val de Travers about 200 km to the north and in the Swiss canton of Neuchâtel. The two deposits were obvious and reasonably well-known. Both mines were based on fractured limestone rock impregnated with about 10% by mass of bitumen. Limestone was a good host for the bituminous intrusions as it was often jointed and fractured and had cavities where water had dissolved limestone and removed any by-products. The fluid bituminous intrusions would have slowly solidified as temperatures dropped and volatiles evaporated.[515]

On the other hand, bituminous material from the Trinidad Asphalt Lake in the south-west corner of the island of Trinidad was a mixture of about 40% bitumen and 60% sand and so was more like a firm mastic than a rock.[517] It could also contain clay, which complicated its use for road-making. The Lake has an area of about 50 Ha and is about 40 m above sea level. Its bitumen is being continuously replenished from underground sources and is probably the world's largest bitumen deposit. The raw material was removed from the Lake surface in large chunks. European sailors had long been aware of the Lake and it gained some English prominence in 1595 when Sir Walter Raleigh stopped there and used bitumen from the Lake to caulk his wooden ships. Norwegian wood pitch was widely used for caulking, but the Trinidad bitumen was preferred as it had a much lower melting point. Before being used in road-making, it is refined by heating to over 100° to remove water, gas and vegetable matter, leaving a mixture containing about 60% bitumen.

An interesting person to now enter our story is Eirini d'Eyrinis (or Eirinus). He is said to have been a native of Russia, a professor of Greek and a doctor of medicine working in Switzerland. His three publications[518] – La Nosourotechnie (1712), Avis sur l'usage d'un Asphalte (1719) and Dissertation sur l'Asphalte ou Ciment Naturel (Paris: P-N. Lottin, 1721) – were the first to present bitumen to the world, technically and commercially. He had visited Val de Travers in 1711 and realised the potential of its asphalt deposits.[519] He approached their owner, the King of Prussia, and obtained a 25-year concession to produce bitumen-based medicinal products, mastics for water-proofing and for caulking ships, and portable fuels.[e] The *asphaltic mastics* were made by crushing the asphaltic rock, heating it to soften the mix, and then adding more bitumen to achieve the required fluidity. The added bitumen would have been obtained from previous brews or by using wood pitch. D'Eyrinis reports using pitch from pine trees.[520]

The product clearly had some merits as steps and stairs built using it were said to be still in excellent condition a century later in 1838.[521] In 1802, a richer source of asphalt was found in the limestone in the Pyrimont/Seyssell quarry but it failed to find commercial use until the whole operation was later taken over by Count de Sassenay in 1832.[522] The Count's business acumen led to some use of powdered rock asphalt for footpaths in Paris, Lyon and Geneva.[523] In particular, it was used in 1829 to construct a footpath at Pont Morand in Lyon, about 100 km west of Seyssel.[524] The surfaces were initially too slippery and this was solved by sprinkling the still-hot surface with sand. The other two problems with using powdered rock asphalt for load carrying were that it became soft when heated and brittle when cold. With experience, the sand content of the material was gradually increased to about 40%. This was also a shrewd commercial decision as sand was much cheaper than powdered asphalt. Powdered rock asphalt still finds some use as a pavement surfacing material. A current German product called Gussasphalt is based on this development path.

Sassenay pioneered using powdered rock asphalt to produce asphalt blocks for paving, ramming the material into moulds. He also took steps to ensure that workmen were properly trained in their application.[525] In 1824, the firm of Pillot and Eyquem in Paris used Seyssel powdered rock asphalt to produce asphalt blocks to pave parts of the Champs Elysée. In 1837, this trial was extended and asphalt blocks were further used on the Place de la Concorde.[526] The aim had been to produce patterned black and white paving so pieces of quartz were added to half the blocks. The trial was initially successful but failed after 6 months in service due to rutting and unevenness. This was attributed to the mastic having a very high bitumen content. Nevertheless, such trials also renewed commercial interest in the Val de Travers asphalts.

In 1838, a trial was conducted in Oxford St near Soho Square in London using square blocks (300 × 300 × 75 mm) made from Val de Travers asphalt (Figure 13.2).[527] None of these trials were successful as, like the powdered rock asphalt footpath trials discussed above, the blocks proved brittle in winter and too soft in summer to withstand the loads applied by horses' hooves and hard-rimmed wheels. A comment attributed[528] to Partiot, then

---

[e] Aspects of this account were disputed by Broome 1963 but we prefer Meyn 1873's account, particularly due to his familiarity with the Prussian deposits mentioned by Broome.

Figure 13.2 The 1838 trial of asphalt bock paving in Oxford St, London, at Soho Square, between Tottenham Court Rd and Soho St. The joints are being sealed with a bituminous mortar.[531]

Director General of Roads in France was that *the joints opened, and the surface of the pavement became rutty and uneven, after only six months use.* The general lack of inter-particle contact in bitumen-rich mixes would also have contributed to their poor performance.

Thus, many asphalt-based pavements failed to live up to expectations and the material gained a poor reputation in Europe. An American study[529] in 1843 observed European practice and observed that the various asphalts then being marketed were *very much the objects of speculation ... some of them fraudulent.*

In 1888, an English engineer in a review paper wrote of conditions in the first half of the 19th century:[530]

> In 1838 a trial pavement was laid in Paris, and the matter was taken up by financiers, not concerned with giving the numerous uses to which it might be applied, attributed to it qualities it never possessed, exaggerated the importance of its virtues, and soundly claimed for it an adaptability to almost every conceivable object in constructive works. Speculation in asphalt became so great as to reach fever heat, panic and collapse followed, and this retarded the proper use of the material for a considerable period.

He was possibly referring to the asphalt block trials on the Place de la Concorde discussed above. Then, in 1850, the French counter-observed[532] that English *engineers too easily accepted all the promises of imprudent speculators and suppliers.*

In France, Léon Malo developed an improved method for making asphalt blocks and, in 1873, de Smedt (Chapter 15) used these blocks with limited success in St Louis in the USA. Compressed asphalt blocks were also used in San Francisco in the 1870s.[533] This field experience and improved machinery led to a further increase in the compression applied during block manufacture and from 1885 the new blocks were providing successful pavements in Baltimore. Two other changes were that broken stone was added to the powdered rock asphalt and the blocks were now 150 mm deep. This was twice the depth of the European blocks – an important change as strength and stiffness tend to increase with depth squared. Asphalt blocks were popular in the early 1900s and continued to be used until the 1960s, finding more favour than they had in Europe. Asphalt block paving placed on Horse Guards Parade in London in the early 1900s is reported to have lasted for 60 years.[534] A common experience was that, to be successful, asphalt block paving had to be placed on a dry concrete basecourse[535] and typical initial block thicknesses of about 40 mm meant that they would not have been self-supporting and were more of a surface course.

In 1840, de Coulaine placed 40 mm mastic layers on a macadam pavement at Saumur in the Bordeaux to Rouen road (RN 138). After a year of successful service the method was extended to other Loire-side towns and to the avenue de Marigny in Paris. The pavements were still performing well in 1850.[536]

The first recorded direct use of asphaltic rock for road pavements was the work of the Swiss engineer André Mérian who, in 1848, had been appointed the chief engineer for the Neuchâtel canton. In 1849, he visited the asphalt quarry in Seyssel where both he and mine manager Léon Malo, a French engineer, had observed that the wheels of the quarry trucks running over rock asphalt chips had serendipitously created excellent in situ paving.[537] Malo later described[538] the event thus: *It is, without doubt, these circumstances which inspired the idea of compressed asphalt highways ... and the Swiss engineer Mérian was the first to profit from this lesson.* Recall that the serendipitous quarry truck story had already arisen with broken stone paving and Tarmac. One of the authors has elsewhere listed many other similar cases related to the discovery of asphalt paving.[539]

Mérian returned to Val de Travers and undertook successful trials of local powdered rock asphalt (*asphalte coulé*) on the road between the mine and Les Verrières. The powdered rock asphalt test pavement lasted at least 25 years. The early success of the trial and a visit to London led Henri Darcy, Inspector General of the national Ponts et Chaussées, in the same year to conduct a similar, partially successful, trial on the Champs Elysée in Paris. [540] In these instances the powdered rock asphalt was rammed into place at ambient temperatures. In an account published in Lyon in 1851, Mérian attributed the success of his asphalt pavements to the fact that the bitumen content did not exceed 15%. Hand ramming and tamping were soon to be supplemented by rolling.[541]

Léon Malo was a member of l'Académie des sciences, belles-lettres et arts of Lyon and became one of the most influential people in the early development of asphalt. In 1866, he published a well-used guide to asphalt paving.[542] Malo had been involved managerially with Seyssel asphalt since 1838 and in the 1850s he had taken over as manager of another Seyssel asphalt quarry, Cavaroche. In 1854, he teamed with H. Vaudrey, a road engineer, to place a mastic of powdered rock asphalt, heated in adjacent cauldrons, onto a small length of Rue Bergère near Rue de Conservatoire de Musique in Paris. The pavement lasted about 60 years.

Following the Rue Bergère success in 1858, some 3,000 m² of paving was constructed on the Rue Faubourg St Honoré and at the Place du Palais Royal. About 50 mm of the powdered rock asphalt was placed on 150 mm of concrete.[543] The very prominent and heavily trafficked locations ensured that the performance of the new paving method was very evident to many in authority. Unfortunately, some parts of the paved area had recently been excavated to a depth of about 4 m for drainage works and it appears that the replacement material had not been well compacted. Consequently the 50 mm mastic of powdered rock asphalt soon cracked. This was a set-back for the local asphalt industry and led observing English engineers to require any asphalt to be placed on a thick concrete basecourse.[544] Nevertheless, the problem caused by the soft subgrade was soon remedied and the new asphalt paving was still performing well in 1880.[545] The use of an underlying layer of concrete had been used by the Romans (Chapter 6) and had been reintroduced in England in the 1840s on the Brixton Road.[546]

The asphalt was about 50 mm thick and compacted and levelled with a heavy roller.[547] Using a compacted layer of powdered rock asphalt had the advantage of producing a surface that was initially water-tight and smooth. However, the water-tightness decreased as the bitumen in the mastic aged and cracked and smoothness was not always a virtue as traffic needed some traction for braking, turning and managing steep hills. These powdered rock asphalt pavements were placed on concrete basecourses, typically about 100 mm thick. The concrete thickness was based largely on prior experience in the local area.[548]

The advent of the railways after 1840 made it much cheaper to bring asphaltic rock to urban areas.[549] Thus it became common to bring pieces of the rock weighing about 25 kg from the quarry to the construction site where they were then broken to about 25 mm in size.[550] These small stones were then ground to sand-like sizes. Crushing asphaltic rock was not an easy process as the final product tended to adhere tenuously to the machinery. The material then had to be heated to about 180°. Various "recipes" were tried, adding small amounts of coal tar, resins or bitumens from previous heatings to the crushed or powdered asphalt in order to reduce the heat needed to produce a ductile mix. Sand was added if the mix was too fluid. The ductility was measured by a form of penetration test (Chapter 11).

In the 1850s in Paris, Darcy and de Coulaine unsuccessfully tried placing the powdered rock asphalt and coal tar cold using a so-called Dufan method which involved using a very fluid mastic.[551] The associated pavements at Place Louis XV near the church of St Roche were initially successful but then suffered severe cracking. In about 1854, it was found that, in order to obtain a successful pavement, immediately before placement, the powdered rock asphalt had to be reheated to between 100° and 120° in large cauldrons at or near the paving site. The heating and melting process could take three to four hours. Additional larger or smaller stones, sand, grit and/or bitumen from various sources were then added to the brew until it appeared satisfactory, particularly with respect to its fluidity. The bitumen content was typically about 15%.

When still molten and well stirred, the mix was brought to the worksite on a small cart, tipped on to the basecourse using large ladles, raked into place, and then compressed by hand-ramming. The surface was finished to a desired smoothness by trowelling it with heavy hot iron sheets attached to the end of long wooden handles (as in the domestic process known as *ironing*). In the absence of heavy rollers, this process would only have consolidated the upper layers of the asphalt.

English engineers had shown little interest in the French asphalts. After a fact-finding visit to London in 1850, Henri Darcy, the French Inspecteur Général des Ponts et Chaussées reported that *the English have little confidence in asphalt. Originally their engineers too easily accepted all the promises of imprudent speculators and suppliers...they (now) reject all new approaches with an explicit but unjustified defiance.*[552]

A commercial advance in asphalt technology occurred in 1855 when Malo and others formed the Compagnie Générale des Asphaltes. In the early 1860s, the Company bought an English rock crushing device called the Carrs Disintegrator and it enabled the much more efficient crushing of rock asphalt to a powdered state.[553] The product was still a mastic as the stone sizes were all quite small. It came to be known in English practice as *compressed rock asphalt* and remained in use as a pavement surfacing layer in thicknesses between 30 and 60 mm until the middle of the 20th century.

In 1869, a Seyssel mastic asphalt was placed over granite block paving in Fifth Avenue in New York. The granite blocks had been noisy and the cause of many accidents involving horses. Within three weeks the asphalt pavement had begun to fail and soon had to be totally removed. The cause was attributed to contractors with inadequate knowledge and experience in the new product.[554] There is also a report of local natural asphalts being used prior to 1860 to pave small areas in Santa Barbara in California.[555]

The early use of rolling was discussed in Chapter 9. Polonceau introduced[556] the first heavy steam-powered rollers in France in 1840[557], where they appear to have been used in the construction of de Coulaine's mastic-on-macadam Loire-side roads discussed earlier. As discussed in Chapter 9, from 1867 practical rollers were being produced by Thomas Aveling using his traction engine production line in Rochester, Kent (Figure 9.8). Aveling had been making the steam-powered traction engines for farm usage since 1860 and his first road roller built in 1865 had used one of his traction engines to tow a 15 tonne, 3 m diameter and 2 m wide cast-iron roller.[558] It quickly demonstrated that the wheels of the traction engine itself would provide adequate rolling, without the need for a towed roller.

The first use of the Aveling steam rollers was in the maintenance of macadam roads, crushing the surficial stones and smoothing the surface. However, steam-powered rolling soon replaced the last two manual stages in asphalt placing and greatly increased the usefulness of asphalt as a pavement surfacing. Rolling with heavy steam rollers had a number of pavement advantages in that rolling could:

- proceed quickly whilst the asphalt was still warm enough to be pushed into its final shape. This allowed pavements to be produced which were properly contoured. The alternative of using straight-edges and hand tamping was too slow, as the cooling asphalt quickly became immovable.
- exert enough vertical pressure to compact underlying broken stone basecourse pavements to refusal.
- apply pressure over wide enough area to ensure uniform compaction-related settlement during construction.
- rectify "soft" spots immediately by rolling in more asphalt.

Thus, rolling would usually produce a smooth and trafficable surface.

The use of a powdered rock asphalt had been strongly promoted in Paris by engineer Baron Quirit de Coulaine.[559,560] Existing cobble and granite block pavements were steadily being replaced by the asphaltic mastics. Its usage in Paris peaked in about 1870 due to the demands of the Franco-Prussian War and then declined due to poor experiences with maintenance and slippery surfaces.[561] An English review in 1876 commented that *the authorities of Paris are about to give up the asphalt paving, and return to the old-fashioned stones, in consequence of the great expense of keeping up the former.*[562] Delano gave a more brutal assessment when he wrote under the heading **Asphalt fiasco in Paris**: *In 1878 the City of Paris let its tried contractor go, tempted by the low prices of a reckless bidder, with the result that in 1883, Paris asphalted streets had gone to the dogs, and had to be relaid in 1884, ... The contractor, who had deceived himself as well as the town, died in a mad-house.*[563] Delano at the time was General Manager of Compagnie Générale des Asphaltes de France Limited. Things were clearly far from perfect. Writing of the period from 1877 to 1883,

Malo commented[564] *the construction and maintenance of asphalt road-ways in Paris had been by an unfortunate decision withdrawn from the (Malo) company which had introduced and acclimatised them in France; how, in leaving its hands, they had fallen by the hard law of lowest tender ... and were finished by a big financial catastrophe.*

Asphalt continued to be used as a surfacing rather than as a complete pavement, with the compacted powdered rock asphalt mastic typically placed as a 50 mm layer on top of a 150 mm layer of concrete. This followed from the Rue de Rivoli experience noted above and had become general practice.[565] Such a pavement profile was used in the first English trial of powdered rock asphalt (from Val de Travers) which occurred in 1869 in London on Threadneedle St near Finch Lane.[566] In this case about 50 mm of asphaltic mastic was supported on 200 mm of concrete. Heavy rain during placement resulted in a further 10 mm of mastic being added. There were mixed views of the performance of the pavement, with one report from 1874 claiming that it had been *most cruel and dangerous to horses*[567] whereas three experts in 1877–1880 reported that the Street was in good condition and had had few repairs.[568] Asphalt made from German rock asphalt was used on Lombard St in London in 1871.[569] The use of mastic asphalt pavements remained a controversial topic in London throughout the 1870s, particularly in relation to damage to horses.[570]

The Seyssel quarries were operated by a French company and it had supplied most of the asphalt being used on French pavements although the asphalt from the Val de Travers quarries remained an attractive alternative due to its higher bitumen content. In 1870, the canton of Neuchâtel gave the concession to operate the Val de Travers quarries to an English company, Neuchatel Asphalte Company Limited. It had been registered in 1873 at the instigation of some Geneva bankers who had realised that a larger capital investment was needed at the Neuchâtel mines. The company subsequently supplied asphalt to Britain and many other countries.[571] The concession was renegotiated many times and the last record has it terminating in 1991. The Company also gained access to Seyssel asphalt. To complicate matters, it had been required by government to allow asphalt in Britain to be marketed by a separate company, Val de Travers Asphalte Company Limited, registered in 1871. For the next 50 years Val de Travers ACL had an effective monopoly on supplying asphalt to English users.

The practice of using powdered rock asphalt mastic as a surfacing (perhaps 50 mm thick) on top of a pavement, rather than as a structural course, was still normal practice in England in 1900 (Figure 13.3).[572]

To this stage we have been discussing pavements made of asphaltic mastics. To add stones to the mastic and thus produce a paving asphalt might have seemed an obvious step, but the mixing and placing of the resulting material would have required powerful machinery. As will be discussed in

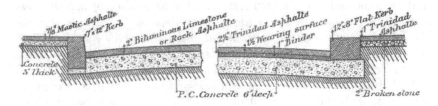

Figure 13.3 Typical English asphalt paving, c1900.[573]

Chapter 15, such machinery was only just beginning to emerge at the beginning of the 20th century.

Although the United States had a number of rock asphalt deposits, the imported bitumen-rich Trinidad Lake asphalt proved a much more attractive raw material, both technically and commercially. This became widely accepted and by 1865 Trinidad was exporting more than 17,000 tons a year, mostly to the United Kingdom, Belgium and France.[574] It was sometimes too hard for direct use and was fluxed with lighter oils.[575]

Bitumen was also used in the first asphalt pavements in two other ways. First, it was often found useful to coat the underlying basecourse with a film of bitumen (a binder layer) to enhance adhesion between the surface course and the basecourse. Second, to reduce the slipperiness of a pavement surface, the asphalt was sometimes brushed with bitumen and then sprinkled with sand whilst the bitumen was still hot. This practice continues to the present day. Depending on the viscosity of the mix, sand in the mix could settle to the bottom of the layer leaving a bitumen-rich upper surface that could be too soft to carry traffic. To overcome this, further sand was added to the mastic surface. In considering these arrangements, it is useful to remember that 19th-century asphalts were actually mastics and no stones larger than 20 mm were added to supplement the sand.[576]

A major step occurred in 1861, when Edward de Smedt, a Belgian chemist, emigrated to work at Columbia University. By 1864, the work led to the commercial exploitation of Trinidad Lake bitumen although de Smedt also tested powdered rock asphalt from Ritchie County in Virginia. A significant application occurred in 1870–1871 when de Smedt mixed heated Trinidad bitumen with sand and powdered limestone to produce a mastic which he used successfully on William St outside City Hall in Newark, New Jersey.

At this time he also began working with Clifford Richardson, another chemist who also had a strong interest in geology, had been with the Department of Agriculture and had travelled extensively in Europe. de Smedt and Richardson realised that the maximum strength of the material was achieved when there was point to point contact between inert stone particles in the mix and no air voids (today such a well-graded mix might be called close-packed or maximum density).[577] This was an advance on mastic and implies that the rock asphalt they studied was only coarsely ground. Trials in 1872–1873 in Battery Park and on Fifth Avenue in New York in front of the Worth Monument used about 40 mm of the new asphalt placed on a 40 mm binder course (see below) which was supported by a 150 mm thick concrete basecourse. The pavements performed well and raised interest in asphalt paving.[578]

In the 1870s, many of the major streets in Washington were in poor condition. This greatly concerned American President Ulysses Grant who had a public desire *to make Washington a capital city worthy of a great nation*, particularly in view of the nation's forthcoming centennial celebrations. Pennsylvania Avenue which runs from the Capitol building to the Whitehouse was the city's major ceremonial street. It had been macadamised in 1832, paved with cobblestones in 1849 and with wooden blocks in 1871. The wooden blocks had rotted within a few years. In 1876, these events led to de Smedt being appointed to a new position of Inspector of Asphalts and Cements within the ambit of the Army Corps of Engineers. In 1877, de Smedt used Trinidad Lake bitumen for paving in a trial to replace rotting wood block paving placed only a few years earlier. The mix was an asphaltic mastic comprising 14% bitumen, 76% sand and 10% pulverised limestone. The trial was a success and became the basis for American *sheet asphalt*. It also boosted the use of Trinidad Lake bitumen in the USA. A companion competitive trial using powdered Val de Travers rock asphalt also lasted a respectable 10 years, but was considered too slippery.[579] By 1886 Washington had about 100 km of sheet asphalt paving. The least successful pavements were those using coal tar rather than bitumen.[580] By 1910, all of Pennsylvania Ave was paved with sheet asphalt using Trinidad bitumen and placed on a concrete base. With minor repairs, the pavements lasted at least 50 years.[581]

In 1887, de Smedt stepped down from his Inspector position and was succeeded by Clifford Richardson, who has been characterised as a *keen observer and shrewd scholar (rather than) a great innovator*.[582] In 1891, Richardson issued new paving specifications[583] which required a 70 mm sheet asphalt course supported by a 150 mm of concrete basecourse. The sheet asphalt surface course was initially regarded as a *cushion* course to even out irregularities in the profile of the basecourse. A 40 mm thick bitumen-rich mastic *binder course* was used to provide a key between it and the basecourse. It did not work very well as it was readily shoved and rutted by traffic. Over time, mix modifications to the original sheet asphalt led to a modified sheet asphalt (or *carpet course*) which avoided the need for the binder course and worked well in most circumstances.[584]

In the 1908 edition of his widely-used text book, Richardson continued to emphasise the common view that the role of the concrete basecourse was to be the supporting foundation that carried the traffic loads and that the role of the asphalt was to protect that basecourse from wear and disintegration.[585] He notes that it was not unusual to use existing macadam, cobble or block paving as a basecourse.

In 1894, Richardson resigned from his government position and became a director of the Barber Company which had been started in the 1870s by Amzi Barber who was a school teacher and University professor who dabbled in Washington property speculation. This hobby had convinced him that street improvements had a major impact on property prices and inspired by the Pennsylvania Ave success, he formed an asphalt paving company. In 1877, his new company had gained the franchise to import the Trinidad Lake bitumen needed for the Pennsylvania Avenue project discussed earlier. In 1888, he convinced the British Government to convert the franchise into a 20-year monopoly covering the entire Lake. As a consequence, and with good reason, Barber became colloquially known as the *Asphalt King* and his associates were called the *asphalt trust*. Barber lost control of the Company in 1901 due to overly ambitious expansion schemes.[586] In the meantime, Richardson's Inspector position had been taken by Alan Dow, who later invented a practical way of assessing the core properties of bitumen and was able to break the trust's monopoly by showing that bitumen from oil refining was at least the equivalent of the Trinidad product.[587]

An 1897 estimate of the amount and type of street paving being used in American cities is given in Table 13.1. Tilson's amended data in the Table is linked to a claim that in 1897, Buffalo, New York State, had more asphalt paving than any city in world, exceeding both Berlin and Paris. Certainly, in 1885, Buffalo had about 30 km of asphalt paving and only Washington had more.[588] Clearly, the momentum for paving had moved from Europe to the United States.

In response to pressure from groups of road users, including cyclists, in 1893, the American government established within its Department of Agriculture an Office of Road Inquiry *to make inquiries in regard to the systems of road management*.[592] It became the Office of Public Road Inquiries in 1899. In 1901, the Office opened a four-storey laboratory for testing road-making materials. It had grown out of a Division of Tests within the Federal Bureau of Chemistry. In 1900, Logan Page, a geologist from Massachusetts who had

Table 13.1 An 1897 survey of amount of street paving in American cities with populations of over 10,000 people[589]

Paving type	Paved length in km	McShane's percentage[590]
Unpaved		45
Granite setts	1,920	10
Wood blocks	1,220	5
Brick blocks	1,160	5
Rock (macadam/gravel)	910	20
Asphalt	230 (or 730[591])	10
Total paved	5,400 (or 5,900)	50

spent some time with the French Laboratoire des Ponts et Chaussées, was recruited by the Department to study road-making materials, and in 1903, he became a regional head of the Office. In 1905, pressure from users of the new internal combustion vehicles led to Office being renamed the Office of Public Roads with Page appointed as its Director.

The appointment of Page led to attention being directed to the selection and application of materials to improve roads in rural areas, leading in turn to the need to promote the best technology for their construction. At the same time the various State roads associations had developed into the National Good Roads Association which also had support from materials suppliers and equipment manufacturers. A manufacturer of agricultural equipment, Horatio Earle, was active in the Association and instrumental in the creation of the Michigan State Highway Commission of which he became Commissioner in 1903. He argued that with modern techniques roads could be built at a reasonable price and to this end developed the "Good Roads Trains" and the "object lesson" road. In 1901, Earle had hitched together the first good-roads train which ran from Chicago to New Orleans. It consisted of a traction engine, a road roller, a sprinkler, dump wagons, and farm wagons, which were loaded with several hundred people riding to the major event, the construction of an object-lesson road.[593] The nine-car train stopped in 16 cities and five States where object-lesson roads were built. This activity was followed in 1913 by models exhibited at various "State Shows" and described as follows:[594]

> The models illustrate standard types of road construction and represent the modern ideas of highway engineers. All of them are built one-twelfth of the full size. With the exception of the brick model, they represent roads with a hardened surface 16 feet wide and with earth shoulders on each side about 6 feet wide . . . . Among the [types of modern construction] are models showing brick, concrete, asphalt block, macadam, sand-clay, gravel, and earth roads. There are other models showing the process of maintenance, resurfacing, and bituminous macadam construction by the mixing, penetration, and prepared-filler methods. One model shows the various methods of draining and strengthening unsuitable bases for road foundations, while another shows a typical method of treating gravel or macadam roads to make them dustless and to prevent their disintegration under automobile traffic.

Earlier in this chapter we saw earlier how the first commercial application of natural rock asphalt occurred in 1802. For the next 140 or so years, the use of asphalt for paving was seriously hindered by three related factors. First, the objectives of a pavement in terms of its useful life and the loads it could carry were far from well defined. Second, there were no tests that could be conducted to determine whether a proposed pavement would satisfy those objectives. Third, the asphalt industry promoted its particular product largely on the basis that it must be superior because it came from their source quarry. Indeed, as the 19th century progressed, the owners of rock asphalt deposits were giving their products patented names and proclaiming them to be inherently superior to all other asphalts.

Rather than concentrating on the source quarry, moving from mastics to using a well-graded mix of inert materials (broken stones and sand) would have made the various powdered rock asphaltic mastics much stronger in compression and therefore more resistant to rutting. It would also have enhanced their ability to operate as stand-alone structural courses as – without large stones – the asphalt was still predominantly a mastic surface course providing a waterproof surface that met the needs of traffic and urban hygiene. In that case, it had to be carried on a stronger and stiffer underlying basecourse, which could be the natural subgrade, a pre-existing macadam pavement, or a cement concrete or bituminous penetration macadam. However, such a move from mastic to asphalt requires purpose-built, powered machinery of the type to be discussed in Chapter 15.

In summary, the three types of asphalt courses available at the turn of the century and prior to the advent of powered machinery for making and placing asphalt, were characterised by Forbes[595] as:

1. an asphaltic mastic fluid enough to be poured onto an underlying stone course and then trowelled into place (easier to achieve with Trinidad bitumen). A trafficable surface course – or *wearing* course – could be produced by applying sand grit to the still soft surface of the asphaltic mastic.
2. a stiffer asphaltic mastic based on crushed rock asphalt, made fluid by heat and perhaps some additional bitumen, spread on the basecourse and then rolled onto the basecourse to compact and smooth the mastic. This was often known as *sheet asphalt* or *bituminous carpet*.
3. a well-graded de Smedt/Richardson mix (see above), spread on the basecourse and then rolled at temperatures of up to 200°C. This approach would often result in a pavement with the basecourse covered by about 70 mm of sheet asphalt.
4. To which we add: penetration macadam, as discussed above. In 1879, it was observed that these penetration-type asphalts could be used to make the entire pavement, but only on *the back streets of a great town, in which facility of cleansing is of primary importance.*[596]

More generally, a comparison of the merits of various pavement types at the end of the 19th century is given in Table 13.2.

Traffic noise due to iron-shod wheels and hooves on stone pavements is a key factor in Table 13.2 and was a major issue within cities. A respected American road expert wrote[597] in the 1870s that *the writings of eminent medical practitioners are full of testimony to the pernicious influence of street noise and din upon the health of the population.* He quotes Professor Fonssagrives, who we met in Chapter 12 discussing the marsh fevers

Table 13.2 Late 19th-century view of the merits and demerits of different
pavement types. Numbers refer to order of merit with 1 being the best.
See also Table 10.1. Based on contemporary reports[600]

Criterion	Pavement type				
	Asphalt	Brick	Granite	Wood	Macadam
Ease of traction	2	3	5	4	1
Foothold	5	4	2	3	1
Cleanliness	1	3	2	5	4
Noiselessness	2	3	5	1	4
Durability	2	3	1	5	4
Construction cost	3	4	2	5	1
Annual maintenance cost	1	3	2	4	5
Impermeability	1	3	2	4	5
Trafficability	2	3	5	1	4
Travel on slopes	4	1	2	3	5
Uniform wear	2	4	1	3	5
Appearance	1	3	4	2	5
Vehicle operating costs	1	2	3	4	5
Baker's "ideal" pavement[601]	1	2	5	3	4

produced by wood paving, ascribing to traffic noise *the prevalence of nervous temperaments and diseases in the large towns.*

In 1885, FV Greene, who was then a paving engineer for Washington DC, examined street pavements in London, Paris and Berlin. He reported that London then had about 20 km of asphalt paving, introduced mainly because its stone paving was too noisy, but that wood paving was more popular despite its short life. In Paris, wood paving was also the most popular. Asphalt had a bad name due to recent failures. Both Paris and London had also found the European asphalts too slippery. Berlin, on the other hand, preferred asphalt over wood which he ascribed to the "very light traffic" in Berlin.[598] Nevertheless, in 1886, Parisian asphalt eminence, Léon Malo quotes his British counterpart, Colonel Haywood, as stating that in London *asphalt is the best facing for roadways.* Malo also commented very favourably on the asphalt pavements recently placed in Berlin.[599]

Ease of traction in Table 13.2, and its opposite, *rolling resistance*, were of great interest to pavement users up until the arrival of the self-powered vehicle at the start of the 20th century. Hauliers and coach-operators were obliged to calculate the number of animals they would need to pull their vehicles along each length of road that they would encounter. Experimental studies were done in France by Morin[602] and England by Telford and Macneill for their work on the Holyhead Road. Some of the results are summarised in Table 13.3 based on the work by Carey in 1914 in England and Baker in 1908 in America. Morin's French data also indicated that

Table 13.3 Rolling resistance for various conditions[604]

Pavement type	Rolling resistance (newton/ton)
Pneumatic tyres on most pavements	
dry weather	180
wet weather	220
Solid rubber tyres	220
Good macadam	
dry weather	260
wet weather	350
Solid iron or steel tyres on most pavements	220
good dry macadam	290
average dry macadam	330
wet macadam	400
dry gravel	400
wet gravel	570
dry earth	330
wet earth	530

vehicles being hauled along stone block and macadam pavements experienced about the same tractive resistance when new, but with time and no maintenance the resistance encountered when travelling over the macadam could be three times as great.

Much of the attention directed at paving in the second half of the 19th century had been directed at better pavement surfaces in urban areas as cities grew in size and national economies began to feed on the fruits of the industrial revolution. Traffic loads were reasonably well controlled and better tyres and suspensions were being produced.[603] Thus, relatively little attention was devoted to the basecourses and subgrades that supported those surfaces. By the end of the 19th century the type of traffic was slowly changing. The change became dramatic as World War I gave birth to myriads of self-powered trucks – although the magnitude of wheel loads altered little, their number and frequency increased exponentially. Most 19th-century pavements could not resist the onslaught. As will be discussed in Chapter 15, pavement priories shifted from the properties of the surface to those of the basecourse.

Chapter 14

A new form of pavement using thin, sprayed bituminous surfacings

As bitumen and tar became widely available towards the end of the 19th century and powered equipment became available at the beginning of the 20th century, the penetration macadam method discussed in Chapter 13 was able to be taken to new levels of usefulness. Recall that the method was to place a layer of clean broken stones – such as macadam – on the subgrade and to then pour or spray a fluid layer of bitumen or tar over the macadam surface. There were three main methods of achieving the required fluidity:

- by heating the bitumen to temperatures up to 220°C,
- by adding more fluid materials (so-called *cutters*) such as kerosene, or
- by emulsifying the bitumen with water, wax or soap (Chapter 11).

It was soon found that, to be sprayed, the bitumen had to be free of the small particles found in bitumen derived from rock asphalt. To achieve full penetration of the stone layer, the fluid bitumen was brushed and squeezed into the surface. However, equipment developments after 1912 made it possible to heat the bitumen at the point of spraying to reduce its viscosity and to then machine-spray it onto the pavement via a transverse line of nozzles.[605] The pavement layer was then rolled before the bitumen had had time to set. Very similar machines were also used for producing the "oiled" pavements discussed in Chapter 13.

The resulting pavement was called *penetration macadam*.[606] The first well-publicised trial of the method was near Bordeaux in France in 1880. A range of commercial variations came on the early 20th-century market with names like Granastik, Lithomac, Cormastik, Ferromac, Pykoten, Rhouben and Aerberli. When it was successfully placed, penetration macadam was similar to the simpler and cheaper water-bound macadam (Chapter 9) but had a much longer service life. The same process using cement mortar rather than bitumen would produce a concrete (Chapter 17).

One problem with the method was that molten bitumen is very hot and sticky and so the process had many practical limitations as the sprayed

bitumen adhered tenuously in many undesired places. Tar was sometimes used in place of bitumen, particularly if the local tars were more
fluid than commercial bitumens.[607] However, it was not widely preferred
as many tars solidified very slowly and then became hard and brittle when
they did finally set. Some of these issues had been resolved by the end of
the 19th century when natural tars were being distilled in order to have
more useful road-making properties. These *road tars* were still in use in the
1960s, after which their carcinogenic properties and a drop in the relative
price of bitumen led the complete replacement of tar by bitumen.

Although the sprayed bitumen did not always fully penetrate the layer of
stones, it still waterproofed the pavement by filling its upper level interstices
with bitumen. This ensured that the pavement and subgrade remained dry
and, therefore, relatively strong and stiff. It had the additional advantages
of suppressing dust and of holding the surface stones in place when they
were subjected by traffic to horizontal traction forces, with each raised
stone being a micro-impediment to forward motion of the pavement surface. A modern development of the method is called a *"spray and chip seal"*
and is discussed further below.

The problems with dust in the era of speeding cars was mentioned in
Chapter 9. In 1902, a Swiss physician, Ernest Guglielminetti, had been
invited to Monte Carlo to give a lecture on mountain sickness. He so
impressed the ruler of Monaco, Prince Albert I, that the prince asked him to
help solve the dust problem on the Grand Corniche road outside the Casino
which was seriously affecting profits from his Casino.

Guglielminetti had seen tar sprayed on hospital floors in Indonesia and
so he developed a method for spraying tar on roads, initially for the road
outside the casino.[608] By careful viscosity-based selection of the tar used,
he solved the slow-setting problem that had plagued earlier efforts and his
Casino trial was very successful. However, there were still many early failures, particularly when the sprayed surface was trafficked by vehicles with
solid tyres. These problems were gradually solved, and the method was subsequently used throughout France where Guglielminetti came to be known
as the Tar Doctor (le docteur Goudron) (Figure 14.1).[609] The process was
subsequently enhanced by further improvements in spraying, particularly
by replacing the tar with bitumen and making it more fluid at ambient temperatures by adding lighter petroleum products such as kerosene and petrol
(called *cutters*) and by emulsifying it with fatty abattoir by-products.[610]

The product as initially used often took over a week to harden in cool
conditions, causing great public distress as it appeared to preferentially
adhere tenaciously to people and private property, rather than to the road
surface. To prevent the resulting surface from becoming too slippery,
bitumen-coated stone chips were spread manually by shovels over the still-
sticky surface. The subsequent loose stones often added to public distress.[611]
Nevertheless, bitumen was better than most of the expensive commercial
alternatives such as *Westrumite*, which was an emulsion of petroleum and

Figure 14.1 Contemporary cartoons of Ernest Guglielminetti, Dr Tar (le docteur Goudron).[613]

ammonia. Many of the alternatives tried irritated the skin and throats of road users and bystanders.[612]

As the above problems were overcome, bitumen or tar spraying of existing or new pavement surfaces, called *surface dressing*, soon became a very effective dust suppressant and a common method of pavement making and maintenance (Figure 14.2). It was soon found that the product was more effective if a layer of coarse sand or small stone chips was spread on and then rolled into the top of the still sticky bitumen (Figure 14.3). When used on an existing asphalt pavement the method was often called *surface

Figure 14.2 Surface spraying in England in the early 1930s.[616]

Figure 14.3 Surface dressing process in England in the 1930s. The construction sequence from right to left is: (1) spraying the bitumen (or tar), (2) spreading a covering of coarse sand or small pieces of broken stone, and then (3) immediately compressing the surface with a heavy roller.[617]

enrichment as the new bitumen replaced bitumen lost from the pavement surface due to wear or ageing. The treatment was far from permanent and it was often necessary to spray road three times a year. Obviously, a major determinant of the long-term effectiveness of the surface dressing was the stability of the underlying pavement surface. If pieces of the surface moved, the sprayed surface was destroyed, at least locally. As quicker setting bitumens came to replace tar, the spraying of surfaces with hot bitumen became more commonplace later in the 20th century. Cement mortars were tried but found to be impractical due to their relatively slow setting times.

In Britain, by 1934 about 60% of its total road length had been surfaced dressed and this proportion remained constant for the rest of the 20th century.[614] One major side-advantage of tar surface dressing was that it was found to greatly increase the life of macadam pavements.[615]

However, major roads had moved to the tarmac-type asphalts to be discussed in Chapter 15 and only 21% of these used surface dressing.

A major development of the approach was the *spray and chip seal* method which was used with great success, particularly for lightly trafficked roads in dry climates and in an era when most speeding vehicles used pneumatic tyres. Thus, it has been mainly used in countries such as Australia, New Zealand and South Africa that had developed their road systems subsequent to the widespread use of pneumatic tyres. It has been most successful when the pavement was constructed while the subgrade was dry.[618] The bituminous seal helps to keep the subgrade dry and is particularly important where the subgrade is comprised of moisture-sensitive materials such as clay and silt.

In the spray and chip seal method, the underlying pavement surface of broken stones is well-compacted prior to spraying, and the sprayed surface is immediately covered with coarse sand and small stone chips. These protect the spray from tyre damage and ensure that surface has sufficient friction to meet the needs of fast vehicles (Chapter 2). While the bitumen is still fluid, the surface is rolled to press the stone covering into the bitumen and ensure that the individual particles interlock.[619] A major development of the method was due to FM Hanson in New Zealand in 1935.[620] His method is based on the amount of bitumen needed to fill the voids in a one-stone thick layer to two-thirds of the thickness of the layer. Hanson was certainly the first to present a measurement-based method for the design and construction of spray and chip seals, and his rational approach remained unchallenged for the next 50 years.[621] The seals usually need replacing every 7–10 years.

Another trend has been to use sprayed bituminous coatings – often using polymer-modified or emulsified bitumens – as waterproofing coats or as tack coats ensuring that two pavement layers stayed in close contact, particularly with respect to forces due to horizontal traction or to expansion induced by temperature differentials between two layers. They also used a surface spray on existing pavements to hold the surface intact, to seal cracks and to ensure that the surface remains waterproof.

Guglielminetti's success with spraying bitumen led other French innovators to experiment with adding natural rubber polymers to bitumen. When granulated rubber became available in the 1930s, Dutch and British investigators conducted further trials. Following the end of World War II, rubberised bitumens were successfully used on Indonesian roads.[622] Late in the 20th century, polymer-modified bitumens were introduced. These commonly were produced by adding liquid rubber to a bitumen. Vulcanised rubber has very long, flexible, strong and cross-linked molecular chains and these can enhance the elasticity and crack-resistance of a bitumen.[623] Thermoplastic synthetic rubbers such as SBS (styrene-butadiene-styrene) and EVA (ethyl-vinyl acetate) have also been used with some success.

Glass fibres have also been used to enhance bitumen properties, particularly to improve its crack-resistance. The fibres are small enough to be part of the bitumen used in a spraying operation. The method is usually much cheaper than the use of a separate geotextile layer (Chapter 11) to manage cracking.

Chapter 15

Asphalt paving produced to carry 20th-century truck traffic

We saw in Chapter 13 that asphalt occurs naturally as rock asphalt. The proportions of the components of rock asphalt rarely led to an asphalt suitable for paving, and much of the 19th-century development of asphalt pavements was hampered by this deficiency and by the lack of sufficiently powerful machinery. Asphalt paving clearly had unrealised potential but was remained very much an art rather than a science. In an address to American municipal leaders in 1904 Alan Dow, a leading asphalt expert, noted that *despite the enormous amount of asphalt paving laid since 1890 in some 120 cities across the nation, prevailing paving methods were almost entirely lacking in system and science.*[624] Asphalt pavements for 20th-century traffic clearly required some new emphases. Common alternatives such as wood block paving had been used with only limited success (Chapter 10).

The increasing availability of bitumen-rich materials from Trinidad showed that bitumen could be used as a paving material in its own right. In Chapter 14, it was observed that early in the 20th century the spraying bitumen on a pavement surface of broken stones could produce admirable pavements meeting most practical operational requirements for moderately trafficked rural pavements. Urban pavements, however, required a more coherent surface. There was also a realisation that good asphalt could be made without using rock asphalt but by mixing Trinidad bitumen, stone and sand.

Manufactured asphalts using bitumen as the mortar in a sand mix producing bituminous mastics were well used in the 19th century. The material was mainly used for surfacing pavements and was commonly called *sheet asphalt*. A thick binder course had been introduced late in the 19th century (Chapter 13) as a key between the surface course and the basecourse. It never met its original objectives and, in the 20th century, it was first replaced by thinner (20 mm) mastic coat and then, by 1915, it had been found that an adequate key could be achieved by simply spraying bitumen (Chapter 14) onto the surface of the basecourse. Moreover, the sprayed bitumen effectively waterproofed the underlying pavement structure.

With the disappearance of the binder course, the name *sheet asphalt* became synonymous with *surface courses* in the USA. Sheet asphalt using the mix proportions developed in the USA produced a relatively smooth surface but in England more angular sand was added to meet local traction needs, and the resulting mix was commonly called *sand asphalt*. These asphalts were initially placed manually with the workers using heated rakes, shovels and tampers.[625] Both sheet asphalt and sand asphalt are at the mastic end of the asphalt spectrum.

Sheet asphalt was used as the surface course for extensive paving of the Victoria Embankment in London from 1906 onwards. It used Trinidad Lake bitumen mixed with well-graded sand and Portland cement as a fine filler and was placed to a thickness of about 25 mm. It performed well and encouraged the wider use of the Trinidad Lake product in competition with the Val de Travers material. The surface course was placed on an asphalt intermediate course, although the maximum stone size used was only 25 mm.[626] It is not known what the underlying basecourse was; however, its location adjacent to the retaining wall of a riverside land reclamation project suggests that it was probably composed of granite and/or sandstone pieces.

The move to bitumen was accentuated in the USA where the local rock asphalts were not as useful as their European counterparts and where there was plentiful supply of bitumen from the Trinidad Lake deposit. However, by 1907, the use of bitumen from oil refining had exceeded the usage of Trinidad Lake bitumen and, by 1915, the Trinidad product had been largely displaced by bitumen from petroleum. This caused a new set of problems as by 1920 competitive American road-makers were specifying 102 different grades of bitumen. Federal action reduced this to nine in 1923.[627]

A new commercial complication arose with the growing use of bitumens made more fluid by diluting them to produce what was commonly called *cutback* bitumen. Common diluents were naphtha, kerosene and various light oils. In 1929, there were 119 grades of cutback bitumen in use in the USA and once again Federal action was needed to reduce this to six.[628]

All this was a harbinger of the demise of asphalts produced by the penetration process, and a reduction in the use of asphaltic mastics and the related 19th-century products are discussed in Chapter 13.[629] Improved machinery after World War I led to even better asphalt mixers. However, the effective mechanical spreading of pre-mixed asphalt did not occur until the 1930s.[630]

Adding stones to the asphaltic mastics produced manufactured asphalt. Hand mixing of the mastics and stones provided a variable and relatively expensive asphaltic pavement. The mechanical equipment– stone crushers, sand and stone dryers, rotary sieving screens, bucket elevators, scales, mixers and screeds – that had been used for making and placing pavements late in the 19th century were all rapidly adapted by the asphalt industry to handle the strong and stiff asphalts required for 20th-century pavements.[631] The product produced by the new equipment came to be called *hot mix asphalt* (HMA). It was first used as paving at the Barber Asphalt paving plant in Long Island, NYC, in 1896[632] and piece was still serving well in 1905. The first public use HMA was in Muskegon City in Michigan in 1902 and this pavement was also performing well after 6 years.

In 1896, the Hetherington and Berner Company of Minneapolis developed portable asphalt plants, mounted on railroad flat cars[633] and this was followed by Barber Asphalt's rail-based plant in 1899.[634] Many consider that the first asphalt production facility to contain most of the basic components of a modern asphalt plant was built in 1901 by Warren Brothers. The Warrens were seven brothers who had commenced business in 1890 in East Cambridge, Massachusetts, as an asphalt roofing firm. Warren Brothers introduced two major changes: the components were preheated and then mixed using purpose-built machines. Heating the stones before they came into contact with the hot bitumen was a major benefit in cold climates. The first asphalt mixer able to be operated at a construction site was introduced in 1909. However, as equipment further developed, asphalt was often manufactured at purpose-built offsite plants and brought to the work site in (often insulated) trucks. The recipe and the mixing process used were critical in the delivery good asphalts. Overheated bitumens would quickly harden and crack in service, and an excess of bitumens and/or diluents would delay the setting of the asphalt and soften the completed product.

Looking now at the specific development of asphalt for 20th-century traffic, we see that as experience with asphalt continued, there was a growing realisation that asphalt could be used not only as a surface course but also as a structural basecourse. A number of mix proportions were tried for basecourses. An early favoured recipe was to replace the particles of sand and fine stone with larger stones and to increase the bitumen content to about 16%. Technically, this produced a gap-graded mix (Chapter 4). This product came to be called *stone-filled sheet asphalt* in the USA and *hot rolled asphalt (HRA)* in Britain. Due to the high bitumen content, it was better in tension and fatigue and therefore less prone to cracking in service and relatively impervious, relying on the cohesive properties of the bitumen. However, the lack of particle to particle contact and the propensity of bitumen soften and flow when heated meant that the product did have some limitations in compression, particularly in hot climates where it could rut or be shoved into ridges under traffic loads. HRA was widely used in Britain, but in the USA, it was mainly used a surfacing layer in thicknesses of between 30 and 100 mm. American asphalts would subsequently tend more towards well-graded rather than gap-graded mixes.

Powerful new equipment was to continue to have major effects. To quote[635] Crawford, Technical Director of the American National Asphalt Paving Association:

> The 1920s and 1930s were decades of great improvement and innovations in mixing facility and laydown operations. The Butler spreader of 1924 gave way to machine laying and finishing in 1928. The first finisher with distributing augers appeared in 1931 and the floating screed 1932. Tow points for screed levelling arms came into use in 1934.

A review of American asphalting practice nominated 1924–1928 as the period during which asphalting changed from *hand labour and animal power* to *mechanical equipment and gasoline power*. Until then, asphalt was placed in a single layer, which made compaction difficult.[636] The spreading and rolling of pre-mixed asphalt in multiple layers became possible with a new plant. [637] These machine improvements led to the current range of asphalts which are stiffer and based more on adding sand to a stone mix, and thus increasing strength and stiffness and minimising the use of expensive bitumen.

It has been suggested that the incorporation of broken stone (increasingly known as coarse aggregate) into mastic mixes to produce a strong manufactured asphalt was a gradual process during the early decades of the 20th century.[638] It was stimulated by the success of mixes of Portland cement, sand, stone and water to produce concrete and was therefore often known as *asphaltic concrete (AC)*. However, in Britain the preferred term was *dense bituminous macadam* or *asphalt macadam*. The bitumen content had to be just sufficient to create the mastic, coat the larger stones and fill any voids as any excess bitumen would destroy interparticle contact and thus soften the end product. An alternative early way of preparing asphaltic concrete was to place a layer of broken stone and then cover it with a layer of bituminous mastic which partially penetrated the stone layer naturally before being forcibly rolled into the layer.

The term *asphaltic concrete* was not new and from about 1860 had been loosely used by some commentators for all paving asphalts. However, by the 1920s in the USA, it was only applied to the new stiff mixes described in the preceding paragraph. An early application of asphaltic concrete was on Long Island (NY) in 1896. It was associated with a decision by New York City to use asphalt paving rather than bricks, granite setts or wood blocks. However, the City also required 15-year warranties on workmanship and materials, and pavement failures caused by factors beyond the asphalt contractor's control bankrupted many builders and raised the cost of new asphalt paving.

Richardson modified his original specification for asphaltic concrete on the basis of in-service experience and he and the Barber Company, with some success, patented the specification in 1903. There was some precedent for this as the first American patent for asphalt was taken out by Nathan Abbott in New York in 1871, based on a tar-based asphalt pavement he had placed in Washington in 1868. Abbott made a number of positive contributions, including better mixing and rolling equipment, which enabled him to place asphalt up to 300 mm in thickness.[639]

In 1905, Richardson published a book[640] on asphalt and its specification, and when Barbers subsequently realised the usefulness of the specification, it is said[641] to have proceeded to buy back as many copies as possible. The main effect of the 1903 patent was to prevent others from making asphaltic concrete with stones larger than 12 mm.

The grading of the stone and sand and the amount and viscosity of the bitumen used in the new asphaltic concretes were carefully selected, based initially on the work of Richardson,[642] to achieve an asphalt which was dense and impervious and had low air voids. Learning from the earlier successful experience with sheet asphalt, this new generation of asphalts was made by using a well-graded stone mix and by dropping the bitumen component from about 16% by mass to 8%. Such a mix would not subsequently compact under traffic; however, the widespread use of steam rollers made it feasible to place and compact such stiff mixes in practical circumstances. Thus, the term asphaltic concrete assumed a far more specific meaning as a well-graded asphalt course, strong in compression, with just sufficient bitumen to fill any interstices and coat the larger stones but not to destroy interparticle contact.

The mix was also cheaper as it required less bitumen and no stone-grinding of the asphalt surface. However, that surface tended to unravel under the action of frequent iron-clad hooves and iron-rimmed wheels and so was found necessary to revert to using a sheet asphalt wearing course until well into the 1920s and the widespread adoption of pneumatic tyres. Initially, an ideal pavement might be a three-layer composite of sheet asphalt, asphaltic concrete and cement concrete. Unfortunately, the competitive nature of the industry meant that things were rarely presented in such a simple fashion. In addition, improved placement methods meant that by 1930 it had become common for pavements to consist of asphaltic concrete placed directly on a 150 mm cement concrete layer.

Prevost Hubbard had headed the American Bureau of Public Roads testing laboratory. In 1919, he joined the asphalt producers' newly created Asphalt Association (later Asphalt Institute). Until the 1920s, asphalts had been selected on the basis of previous empirical experience with mix proportions and components. In the 1920s, Hubbard and Frederick Field introduce a test to detect the rutting propensity of asphalts, particularly sheet asphalts. A 50 mm diameter specimen was subject to a plunger test using a hand rammer. Tests of this nature were unfortunately called "stability" tests, although they measured resistance to deformation rather than a change of state implied by "stability". Thus, they were primarily directed at the propensity of the asphalt to be shoved horizontally or rutted vertically.[643] The 50 mm sample size meant that they were also inappropriate for any asphalts which used large stones. Nevertheless, the test led in the USA to the widely used Hubbard Field Method of asphalt design. Its main virtue was that it was simple to apply to mix design, and its use persisted in some areas until the 1970s.[644]

By the 1960s, the test had been modified to use larger specimens, six different hammering processes and a separate compression loading of the specimen, finishing with the specimen loaded when immersed in cold water. In the late 1930s, Bruce Marshall at the Mississippi Department of Highways had developed a further version of the process. It was lighter and more portable than the relatively heavy Hubbard-Field device and so came to be widely used for airfield construction during World War II. A further development of this approach was produced by Francis Hveem in California and was popular in the USA in the 1950s.[645]

The Hubbard-Field, Marshall and Hveem tests did not explore any causative parameters in an asphalt and failure in a test bore little relevance as to how asphalt failed in service. The tests were useful as a means of ensuring consistency between asphalt mixes but, as they did not relate to in-service performance, they had no relevance beyond the empirical boundaries within which they had been calibrated. They were also relatively time-consuming and expensive. Gyratory tests developed in the 1950s and standardised in the 1970s better represented the way asphalt is compacted during pavement construction.

The slipperiness of asphalt surfaces was a major user concern. The sprinkling of sand on the surface and the pressing of broken stones into the surface have already been mentioned. On busy roads carrying horse traffic, it was often found necessary to wash mud and refuse from the surface on a daily basis.[646] The surface of a new asphalt pavement was often roughened by rolling precoated stone chips into the still-warm asphalt surface.[647] The technique may have been first used in Surrey in England in 1928. A further problem arose with bitumen-rich mixes such as the British hot-rolled asphalt (HRA) where the pressure from the wheels of traffic, particularly in hot weather when pavement temperatures can reach 60°C, can press the more fluid bitumen to the surface (called *bleeding*) leaving long smooth bitumen strips in the wheelpaths (Figure 15.1). The strips were often not perceived by drivers and the loss of traction was a common cause of crashes. A variety of methods were then developed to measure pavement surface friction (Chapter 19) and alert pavement managers to the need to take remedial measures.[648]

Cracking can also be a problem with asphalt pavements. Mixtures of bitumen emulsions and some sand came to be used as *slurry seals* to fill cracks in and make minor shape corrections to existing pavements surfaces. They work because emulsions are mixtures of bitumen and water which

Figure 15.1 Wheelpath "bleeding" of a pavement.[649]

are initially fluid at ambient temperatures when placed in a crack but then quickly break down to their separate components. Simple coatings of bitumen have also been often used as a sealing and waterproofing coat.

The development of asphalt paving in Britain at the beginning of the 20th century was influenced by a number of factors. First, local firms like Tarmac had developed a strong domestic foothold with its specific slag-based product (Chapter 18). Second, there was considerable interest in the successes and failures of asphalt paving in Paris. Third, there were strong commercial links between British and European asphalt interests. Fourth, some influential Englishmen had growing interest in promoting the use of bituminous materials from the Trinidad Lake deposit.[650]

The British development of the bitumen-rich hot-rolled asphalt was discussed above. An alternative product of the new mixers was called *coated macadam* and had its origins in the sheet asphalts produced using ground rock asphalt. A well-known but unsatisfying British definition of coated macadam was *a road material of graded aggregate that has been coated with binder and in which the intimate interlocking of the particles of aggregate is a major factor.*[651] *Graded aggregate* was misleading as coated macadams were characterised by air voids which were large and often interconnected. In the pavement industry *aggregate* is a common term for stones. As with the original macadam, coated macadams obtained their strength from interlock between the stone particles, whereas hot-rolled asphalts placed more reliance on the mortar for their strength.[652] Coated macadams were cheaper than hot-rolled asphalt and easier to place, but they were relatively permeable and could compact under traffic.[653]

When the binder used was tar, the product was called *tarmacadam*. This created some confusion as *tarmac* was a proprietary product using slag for its stones (Chapter 13). Tarmacadam was the major paving material used on English main roads during the 1930s.[654] The confusion was heightened by the fact that, in the 1930s in England, there were seven different Standard Specifications for hot-rolled asphalt alone.[655]

An even greater profusion of terms occurred in America where, in the first decades of the 20th century, a variety of proprietary and patented asphalts were being marketed with names such as Willite, Romanite and Bitulithic. This led Hveem from the Californian highways agency to comment wisely[656] *materials all depend on the same principles for successful performance. They do not care what they are called by promoters, salesmen or engineers – they will all continue to act and respond to the same natural laws.*

The Warren Company had broken away from Barbers in 1900. In response to the major rutting problems with sheet asphalt pavements, Warren introduced and patented an asphaltic concrete which they called Bitulithic.[657] It was based on tests by Warren which showed the importance of minimising air voids in the stone mix used as filling voids with expensive bitumen not only increased the cost of the asphalt but the resulting pockets of bitumen also softened the asphalt. The patent covered stone sizes of up to 75 mm and so could accommodate vertical loads via point-to-point contact between the stones. Nevertheless, the first Bitulithic pavement – placed in Rhode Island in 1901 – used stones of 20 mm or less. This was probably in response to tendency of iron-rimmed wheels to crack any surface stones and the related difficulty of producing a smooth surface with a mix containing large stones. However, better equipment and a growing fleet of rubber-tyred vehicles allowed Bitulithic to capitalise on a much less demanding set of surface requirements. Sheet asphalts also became more relevant in the 1920s as pneumatic rubber tyres became the norm. The development of more efficient mixing and placing equipment allowed stiffer asphalts to be produced by using bigger stones.

On the other hand, a patent decision in 1910 (The Topeka Decree) determined that mixes using stones smaller than 12 mm did not infringe the Bitulthic patent. These small stone sized AC mixes were rarely employed in countries not covered by American patents.[658] The patents for Bitulithic pavement expired in 1920, and subsequent improvements in pavements by Federal and State engineers forced most of the remaining patented pavements from the American market.

Commercial interests often dominated the paving market and led Page to make this statement in November 1911:

> The entire movement for better roads should be so systematized and everywhere placed on so high a plane of honest and earnest effort that the cheap charlatanism of the professional promoter and the bungling efforts of the well-meaning but uninformed citizen should be no longer permitted.[659]

By the 1920s, asphalt was the dominant American paving material. A 1926 survey of the pavement types used in the 12 largest American cities gave the following results:[660]

asphalt	55%
stone-blocks	15%
water-bound macadam	13%
brick	13%
concrete	3%
wood block	2%

Table 15.1 summarises the role of the various components of the current range of paving asphalts. *Filling voids* is a key feature of the table as minimising the void content of an asphalt mix – typically to about 4% – has remained one of the most important facets of asphalt production. Minimising voids increases rutting resistance, fatigue strength, stiffness and impermeability.

Large, poorly maintained and/or badly operated, trucks can be a major source of traffic noise, but in modern traffic conditions, most of the annoying noise is generated at the tyre-pavement interface during braking and accelerating. The amount of noise generated also increases with a vehicle's speed. The search for pavements with good friction properties for braking and turning while generating only low levels of tyre-road noise at speed led to the development for surface asphalt of *open-graded friction courses (OGFC)*, sometimes called *porous friction courses*. These only use large stones and only enough bitumen to coat the stones and not to fill the voids created by the open-grading (Chapter 4). The voids can be up to 25% of the volume of the asphalt. The surface voids absorb much of the tyre noise. They also allow surface water, which can cause unsafe driving conditions, to be removed by draining into the pavement surface, from whence it is subsequently removed through further voids at the longitudinal edges of the surface course.[661]

Table 15.1 Components of asphalt and the consequential asphalt types

Decreasing component size		
large stones	small stones and sand	Bitumen
Provide strength: shear by interlock & compression by stone-to-stone contact	Fill voids between large stones & stiffens the bitumen	Fills remaining voids and glues stones together
Well-graded mix, producing *asphaltic concrete, dense bituminous macadam, or asphalt macadam*		
	Gap-graded mix, producing *hot-rolled asphalt*	
	Mastic, sheet asphalt	
Stone mastic	Stone mastic, slurry seals	
Bituminous macadam & porous course		friction course
Unbound course		

The friction course is placed on top of a conventional asphalt course and is not assumed to be a structural course. Such courses can last between 5 and 15 years. They are effective but are all relatively expensive as their strength comes solely from interlocking of the stones, so they usually require the use of fibres and higher quality stones and bitumen to enable them to maintain their role under traffic.[662] High skid resistance also increases the noise generated by vehicles using the road but some of this noise is dissipated in the interstices in the pavement surface.

Stone mastic asphalt is a mix which only uses large stones and is thus a gap-graded mix. The bitumen coats the stones and fills all the voids between them. Fibres are added to the bitumen to stiffen it and prevent it flowing out of the mix during construction and under traffic. It was developed in Germany in the late 1960s to resist damage from the studded tyres which at that time were still permitted for use on icy roads. It is now used for its rut resistance as the stone-to-stone contact gives it good stiffness, even in high ambient temperatures. To enhance its friction properties, it is sometimes produced by placing a layer of bitumen and then rolling the stones into the layer. The high bitumen content makes it costlier than conventional asphalts.

As we have seen, the basic processes involved in making asphalt pavements involve the following steps:

1. stones, sand and bitumen are mixed together to produce asphalt – the stones and sand will have been dried and possibly heated, and the bitumen will have been heated to much higher than ambient temperatures and possibly also emulsified (Chapter 18),
2. the resulting hot asphalt is placed on the supporting pavement layer,
3. the asphalt is spread to approximately the required thickness,
4. vibrating and floating screeds ensure that the pavement has the right profile and
5. the asphalt is compacted by tampers or rollers. Parts of this step may occur in conjunction with step (4).

Perhaps the major change during the 20th century was the mechanisation of most aspects of this road-making process. As confidence grew, asphalt layers of about 75 mm in thickness were being used as the basecourse on main roads, often without any surface course and replacing the traditional cement concrete basecourse.

The initial wave of big paving machines used tractor treads similar to bulldozers to enable them to negotiate a range of subgrades, but these were gradually replaced with large pneumatic tyres which permitted faster paving and provided smoother finished surfaces.

The vibrating roller compactor had a beneficial influence on both asphaltic and unbound pavements as they allow more compaction to be achieved without increasing the static weight of the roller. They came into practical use in the 1960s. The use of machines for spreading and finishing asphalt was introduced in the USA in the late 1920s, beginning with a Barber-Greene machine.[663] Automatic control of the screeds used to level the finished asphalt was introduced in 1927 in Orange County, California. The next major step was the *floating screed* adopted in 1930, which eliminated any need for side formwork. However, it came into real prominence in the 1960s when the screeds could operate over two pavement lanes. Curiously, much of the machinery used was based on plant previously developed for constructing the rival concrete roads.[664]

The processes were further enhanced by the development during World War II of better facilities for producing hot-mixed asphalt. Processes for allowing a continuous flow of asphalt paving along a roadway were developed in the 1980s. These culminated in the *autograde* machines placing asphalt precisely and efficiently, followed by vibrating rollers delivering the required compaction level. An associated advance was the use of automatic levelling in which the screed follows an optical survey line and adjusts if the placed asphalt surface would deviate from the required level.[665] More recently, GPS technology has added to the accuracy of asphalt placement. Paving equipment has also embraced "intelligent compaction" which uses vibratory rollers equipped with integrated measurement systems, an onboard computer-based reporting system, GPS-based mapping and feedback control. It also provides a continuous record of the properties of the asphalt being placed.

Many of the earlier advances in asphalt paving machinery have since been subsumed in the current range of asphalt paving machines which receive asphalt, place it, compact it and level it to specification and in the required position in a single pass (Figure 15.2). A machine designed for local streets is shown in Figure 15.3 and a large modern asphalt paving machine is shown in Figure 15.4. These machines use automatic control of all relevant parameters, particularly pavement thickness, during the paving process.

A range of bitumens were also becoming available from a burgeoning number of new oil refineries. In recent times, the production of bitumen has become more controlled, particularly as the properties most appropriate to the performance of bitumen in pavements have become better known.[666] Bitumen ductility has also been enhanced by the addition of materials such as natural rubber. The method was initiated by Swiss engineers and was first applied in Sheffield in England in 1932.[667]

Many new pavement asphalts had been tried during the 19th and early 20th centuries as more powerful equipment became available. However, the processes for the selection of the pavement materials and their geometric arrangement were very empirical as there were as yet no quantitative evaluation of the loads that the pavements would be required to carry, of the relevant quantitative properties of the subgrade and of the proposed

Figure 15.2 Modern medium-sized asphalt paving machine – the AMMAN AFT 600-3 tracked machine is moving from left to right.

Figure 15.3 Modern medium-sized asphalt paving machine – the AMMAN AFT 700-3 tracked machine is moving from right to left.

Figure 15.4 Large Wirtgen asphalt paving machine. The paver uses a high-precision site-based GPS which transmits contour data to the machine and collects as-built data from the finished surface. The paver has a mast for receiving data, digital slope sensor for cross fall, grade sensor for slope and an onboard system for controlling the position of the machine and of its screed extensions.[668]

pavement materials, and only the simplest means of assessing how those loads would be distributed throughout the pavement and the subgrade. Proposed pavements were assessed on the basis of the prior performance of similar pavements. The only real paradigm shifts from prior practical experience were McAdam's reliance on broken stones and the use of bitumen as a glue in association with those broken stones. There was also a century-old perception that asphalt was a material for surfacing and had no structural merit. In 1880, William Delano, who had placed many asphalt pavements in Paris on the standard concrete base, observed that *Asphalt in itself has no more power of resistance than sheet lead or India-rubber; therefore it must yield unless well supported from beneath.*[669] The idea that asphalt might be used as a load-carrying and load-distributing structural course arose slowly.

We return to these issues in the next chapter.

Chapter 16

The design of asphalt pavements – a predominantly American initiative

Pavement design is based on a knowledge of the traffic likely to use a road, on the properties of the intended natural formation and subgrade, on the materials and methods that will be available for pavement construction, and on the overall and local climate. These will lead to considerations of the number and type of layers to be used in the pavement and on the road-side drainage. The surface of the pavement also has a major impact on user safety and these matters are covered in Chapter 19. As mentioned at the beginning of the book, the width, slope and general alignment of the pavement are matters of road design. They are each beyond our scope and are well covered elsewhere.[670]

The manner in which the pavement and subgrade are protected from water was broadly covered in Chapter 14. Once this has been resolved, the determination of the thickness of a pavement would ideally and predominantly be a matter of stress and strain analysis based on the principles of structural mechanics. However, as explained in Chapter 13, until the 1930s the selection of the thickness of asphalt pavements was empirically based on experience and experiment. This chapter will cover the progression of pavement design from such empiricism to the current performance-based methods.

A major step towards the structural design of pavements followed from the investigation of a series of Californian pavement failures in 1928–1929. After eliminating failures due to poor compaction and/or drainage, the data highlighted the need to know subgrade properties before selecting asphalt pavement thicknesses. It also indicated that subgrade properties could not be predicted solely from a knowledge of soil type.[671]

In 1930, this led to the California Division of Highways introducing a test – called the *California Bearing Ratio (CBR)* – to assess the structural usefulness of a subgrade.[672] The test was based on a device called the *Bearing Power Determinator* that had been developed by the American Bureau of Public Roads in the early 1920s. The CBR measures the vertical deformation of the subgrade under a static load applied by a flat circular disc (Figure 16.1). The device test samples brought back from the field and placed in a cylindrical mould, hopefully without destroying the relevance of

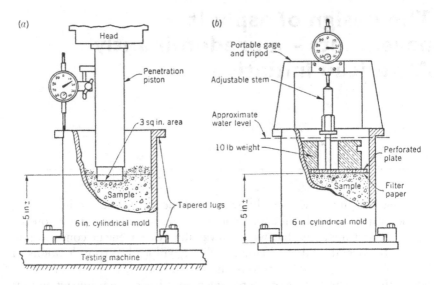

Figure 16.1 The original CBR device.[673] (a) Penetration test. (b) Expansion test.

the data to the field conditions. Onsite versions of the test were subsequently developed. The CBR number was defined as the ratio of the penetration force needed to produce a specified vertical deformation (e.g. 5 mm) to the force need to produce the same deformation in a broken-stone macadam pavement (Chapter 9). The CBR is thus a measure of the stiffness of the subgrade under vertical load.

Californian observations linking performance of practical pavements to pavement thickness and measured subgrade CBR during the 1930s then led to the publication in 1942 of design curves giving the appropriate pavement thickness for a subgrade with a known CBR[674] (Figure 16.2). It was the first time that pavement thickness had been numerically linked to subgrade properties. Note that the minimum recommended pavement thickness was 75 mm. The design process is inherently empirical as the required pavement thickness is based on the subgrade properties given by its CBR value, and the result is not dependent on any properties of the pavement other than its thickness.

The curves were widely used by the American Corps of Engineers building pavements for roads and airfields during World War II. The impetus for this came when the new B-17 Flying Fortresses with very high wheel loads first landed on Pacific airfields. Asphalt and concrete runways based on the existing empirical methods were soon severely cracked or rutted. O. J. Porter from California was brought in to find a solution and soon found that pavements with thicknesses determined by their CBR used with curves such as Figure 16.2 would be much more adequate than ones based on then-current asphalt and concrete pavement design practice.[675]

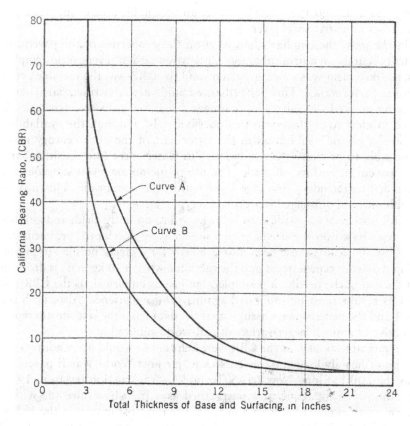

Figure 16.2 The first CBR-based pavement design chart.[678] Curve B was based on empirical field experience, and curve A was the conservative design recommendation.

It can be seen from Figure 16.2 that CBRs of about 4 or less require much thicker pavements, and common practice is now to enhance any subgrade with a CBR of 6 or less by adding another course (or *working platform*) to raise the CBR. It is then a matter of minimising the combined cost of the working platform and the thinner basecourse. At the other extreme, as a consequence of the CBR definition, subgrades and pavements of well-graded broken stones, asphalt or concrete would be expected to have a CBR of 100. With some local modifications, the CBR approach was widely adopted internationally.[676] Nevertheless, in 1955, a British highway expert was suggesting that 100 mm of asphalt would be suitable *if there was any doubt as to the strength of the foundation*.[677]

The CBR is a purely empirical test and sheds no light on how a pavement actually behaves. The implication of the test not taking any account of the properties of the pavement layer is that the role of pavement thicknesses is merely to distribute the wheel load over an acceptable area of subgrade and not to act as a structural layer.

In the past, the emphasis had been on the properties of the pavement, seated securely on a firm subgrade. CBR-based design reversed this emphasis, the pavement was now the "given" and the CBR was the variable determining performance. This performance could vary over time, particularly if water entered the subgrade. A range of material tests were then developed to alert road-makers to this possibility. In addition, the availability of strong plastic membranes in the latter half of the 20th century made it possible to provide effective barriers to water entry into sensitive parts of a pavement and its subgrade. The use of membranes was subsequently extended to include a drainage layer within the membrane. This mix of membrane and drainage layer proved very effective.

CBR testing of subgrades can be awkward to do in the field, and alternative tests based on the needle *penetrometer* were developed, primarily for field use. Basically, these tests measure how far a steel "needle" (typically 10 mm diameter) penetrates into the subgrade when a 10 kg mass is dropped on the end of the needle. It is simpler, but just as empirical, as the CBR test and its results must be calibrated against prior experience. Also, both the CBR and the penetrometer assume that the data from the test site are representative of a much larger variety of real-world subgrades.

Empirical tests such as the CBR were helpful but would never address the design of heavily loaded pavements. A major post-World War II pavement was the 200 km long New Jersey Turnpike. Construction began in 1950, and the pavement required was specified as a frost-free subgrade with a pavement thickness of at least a metre above the frost line.[679] In the 1950s, it can fairly be said that pavement selection and design was still more based on past experiences rather than on structural analysis and the testing of materials. The main design and construction response to the prospect of load increases was to increase the pavement thickness.

Asphalt pavement design based on structural analysis began to occur in the early 1960s.[680] There had already been progress in evaluating material properties and environmental factors such as water and temperature. The main usage-related factors to be considered were the vertical and horizontal forces produced by construction equipment building the pavement and then by the vehicles using the pavement.

Considering construction equipment, as pavement construction proceeds from the bottom up, the subgrade and then each pavement layer and course must be able to provide adequate support for the next round of construction activities such as further material placement, layer compaction and surface rolling.

While braking and turning produce horizontal loads which can damage a pavement surface, the main damage is usually caused by vertical wheel loads acting through the tyres onto the pavement. Note that the weight of the vehicle and its cargo is important only to the extent that it determines the load on each wheel within the truck suspension configuration. This wheel load works through the tyre to become a pressure on the pavement surface. That pressure must be close to the tyre inflation pressure and so the contact area is approximately the wheel load divided by the tyre pressure. In fact, the pressure is slightly higher where the tyre wall is nearest to the pavement.

This contact pressure becomes the stress at the pavement surface. It diminishes in magnitude from the surface to the subgrade as the load is dispersed through the pavement structure (Figure 16.3). In material terms, it is this stress distribution which will determine how the pavement deforms under load. Two types of effect occur. High compressive stress levels from one passage of a loaded wheel may cause permanent compressive deformation of the pavement, and after a series of such passes, the deformation may accumulate and produce a rut at the pavement surface.

The second effect occurs as consequence of tensile stresses occurring as the pavement bends (or flexes) during the passage of a wheel. As each wheel passes, the stress levels cycle between minimum and maximum values and this can lead to tensile fatigue cracking within the pavement (Figure 16.4). Fatigue cracking can reduce the strength of the pavement, reduce the durability of the pavement surface and can allow water to enter the pavement and damage it and the subgrade. As with most materials, fatigue damage depends on the logarithm of the number of load passes – thus 100 passes might only cause twice as much fatigue damage as ten passes.[681]

Given these two effects, an important first step towards rational pavement design was a better assessment of the loads caused by passing traffic. Road vehicles can have a range of wheel configurations and an important pavement design concept that arose was the *equivalent standard axle* (Figure 16.5). This was based on the maximum legal load permitted on an axle in the local jurisdiction. This will be a truck axle rather than a car axle as car tyres carry about half the load of truck tyres. Overloaded wheels

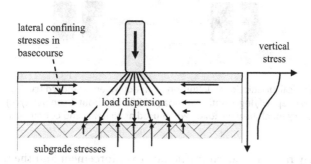

Figure 16.3 Load dispersion in a pavement. The corresponding internal vertical distribution of compressive stresses is sometimes called a *pressure bulb*.

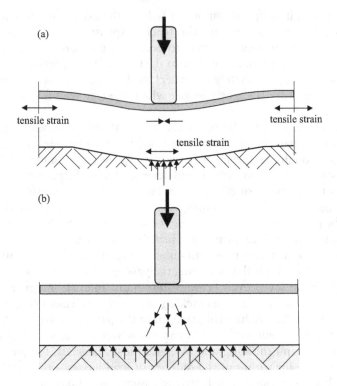

Figure 16.4 Tensile loading strains within a pavement due to bending of the pavement. (a) Flexible pavement. (b) Rigid pavement.

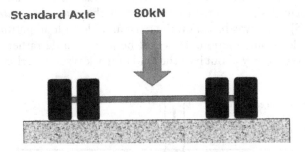

Figure 16.5 Typical jurisdictional definition of an equivalent standard truck axle with dual tyres specifying a stationary load of 80 kN (about 2 ton per tyre). Pavements carry much heavier loads when used at airfields and container terminals.

are relatively rare in modern traffic due to enforcement and the inability of most tyres to carry overloads without causing rapid tyre damage.[682] Local regulations and vehicle configurations are assumed to prevent pavement failure due to the passage of a single "rogue" overloaded vehicle.

Design for a particular traffic situation is based on the number of passes of an equivalent standard axle which would do the same pavement damage as one pass of an axle of an expected vehicle.[683] The emphasis on equivalent standard axles arose because well designed and constructed pavements in regimes where legally overloaded trucks were rare, failed by fatigue after many thousands of truck passes rather than by occasional overload. By the end of the 20th century, empirical pavement design curves such as Figure 16.2 had been extended to account for the number of equivalent standard axles, N_{esad}, to which the pavement would be subjected. Such a chart is shown in Figure 16.6. The fact that the required thickness relates to logarithmically to the number of traffic passes, and not linearly, is consistent with most other cases of material fatigue.

A major step forward occurred with the AASHO Road Test in the late 1950s which involved testing full-scale pavements under real truck traffic.[684] The tests are discussed in more detail in Chapters 19 and 20. One key finding was that the then-current thin asphalt layers were ineffective under heavy traffic. This led the American asphalt industry to move quickly to promote pavements constructed solely of a thick asphalt layer, avoiding the use of a concrete lower layer which had been common practice for the previous hundred or so years. Pavements comprised of an asphalt layers placed directly on the subgrade are now commonly called *full depth*

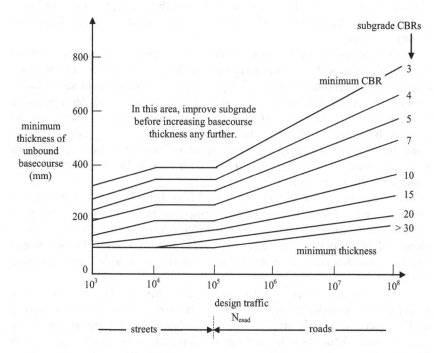

Figure 16.6 Typical empirical pavement design chart, relating pavement thickness to the equivalent number of standard axles in the expected traffic, N_{esad}.[687]

asphalt pavements. Pavements where the asphalt layers are placed on a layer of broken stones or stabilised material (Chapter 18) which in turn rest on the subgrade are commonly called *deep lift asphalt pavements* or, more positively, deep strength asphalt pavements.[685] The AASHO Road Test also produced a pavement design manual[686] which has subsequently dominated American pavement design. Major updates – still based largely on the Road Test results – were issued in 1972 and 1993.

Methods for calculating the stresses and strains in a pavement developed rapidly in the last third of the 20th century, building on the earlier theoretical work of Joseph Boussinesq and others for concrete pavements (Chapter 17) and aided greatly by the advent of high-speed computing. The first of these was probably the Californian ELSYM5 program issued by FHWA in 1972.[688] A key asset of the new methods was the ability to tackle multi-layer pavements. A widely used current program is Mincad's CIRCLY program which is now in its seventh issue.[689] The Shell Company – motivated by its links to bitumen production – produced a program that could accommodate 10 pavement layers in 1978.[690] There are about a dozen computer-based asphalt pavement design programs now in use around the world, based on semiempirical estimates of pavement performance. For a typical input of traffic loading, subgrade properties and pavement layer properties, they will suggest the required thicknesses of the pavement layers. For the design of a rehabilitating layer for an existing pavement, the input will also include a measure of the deflection of the existing pavement.

As methods of stress analysis continued to improve, the pavements were treated as loaded slabs. This required a knowledge of their mechanical properties in tension, compression and shear and the response of the subgrade to applied intermittent dynamic vertical loads. In particular, deflection under load and the distribution of stresses are determined by the stiffness of the asphalt. It became evident by the 1980s[691] that for many practical conditions the dynamic stresses and strains generated within common pavements by traffic loading could be calculated with reasonable accuracy using a multi-layer elastic model.[692]

More recently, finite element models (FEM) have been developed. FEM involves the modelling of a structure's response to loads through the partitioning of a complex system into simpler components called finite elements. FEM models are particularly useful for predicting the non-linear response and deformation of granular pavements to a wide range of traffic loadings.[693]

If overloading is controlled, the two main failure modes are rutting due to deformation under repetitive compressive stresses and fatigue cracking under repetitive tensile stresses. Thus, sophisticated laboratory methods were developed for measuring these mechanical properties of asphalt, as illustrated by the beam bending test in Figure 16.7. To obtain fatigue data,

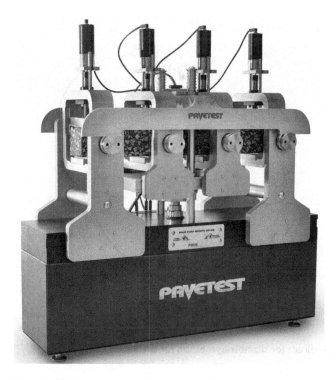

Figure 16.7 Test equipment for measuring the static and fatigue strength of an asphalt specimen loaded as a beam with two intra-span loads producing constant bending stresses in the central third of the loaded beam.[694]

asphalt samples are tested under cyclic load to establish a classic material fatigue relationship linking the tensile strain in the asphalt pavement to the number of load cycles before significant fatigue cracking occurs in the asphalt. While testing for elastic, viscoelastic (time-dependent) and plastic (overload) behaviour is relatively straightforward, such fatigue testing is more time consuming, and the results are more scattered.

The testing arrangement, as shown in Figure 16.7, is relatively complicated, and alternative method has been developed which involves loading a common asphalt test cylinder along a diametrical plane (Figure 16.8) and using the tensile stresses induced at right angles to the loading direction.

The 1990s development of testing equipment capable of measuring all the relevant mechanical properties of asphalt made it possible to test asphalt mixes prior to construction and, if the test results did not meet the design requirements, to then modify the mixes until a satisfactory outcome was achieved. However, given the "local" nature of the materials used in pavements, relevant property data are not always easily obtained. As will be discussed in Chapter 19, this led to the development of methods for

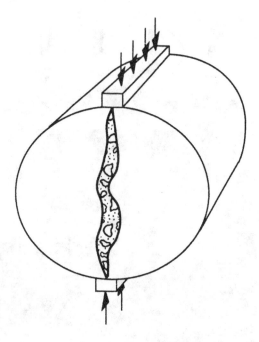

Figure 16.8 Indirect (or diametral) method for producing tensile stress.

retrospectively measuring overall pavement response; particularly deflection response to vertical loading.

So far, we have been discussing methods of pavement analysis and material testing. Design is a different matter. As noted above, prior to the 1980s, methods for the design and construction of pavements were based on experience, empirical processes and material tests directed more at classification than at performance. A reliance on the CBR test had indirectly led to a focus on rutting as the primary performance measure. As understandings of asphalt behaviour improved the methods of design described above – the so-called "mechanistic" methods – gained more prominence. From a material viewpoint, these assume that the stones used in an asphalt mix are of adequate strength to carry the compressive and shearing loads produced by traffic, that the bitumen content is sufficient to do no more than thinly coat the stones and fill the voids between them, and that the mixing and placing processes ensure that this coating and filling regime is achieved in the finished asphalt layer. In this context, using testing arrangements such as that shown in Figures 16.7 and 16.8 made it possible to test asphalt properties to determine asphalt stiffness, creep resistance and fatigue strength.

Having fatigue data was a major step forward. The asphalt thickness is chosen to keep horizontal tensile strains below the fatigue limit. This was in contrast to equivalent earlier design methods which had been based on restricting the permanent vertical deformation of the pavement. For soft subgrades (CBR < 10), similar fatigue-type relations were developed to restrict subgrade deformation.[695] Fatigue-type testing has also made it possible to determine how much an asphalt might compact – and thus rut – under traffic. This compaction is mainly a consequence of a reduction in the air voids in a mix.

In 1948, it was wisely observed that:[696]

> the mathematical approach seems to have a special appeal to those who are able to handle complicated mathematical relations with ease and facility......it is customary to regard soil as an idealised uniform substance......Unfortunately, most real soils are not monotonously uniform and the degree of compaction and moisture content are variable......Vehicle load applications are fleeting.

Despite this warning, the actual pavement design became more empirical and more complex, as the following chart "explaining" the 1993 American (AASHTO) design method demonstrates (Figure 16.9). Its tortuous nature belies its empirical foundations. If this comment seems a little harsh, it reflects a not uncommon view of American pavement research during the 20th century. For instance, after Seely conducted a comprehensive review of the highway research program conducted by the American Government's Bureau of Public Roads (BPR) during most of the 20th century he wrote:[697]

> BPR investigators encountered serious difficulties in producing information of direct utility to engineers in the field. The evidence indicates that the BPR's infatuation with attitudes and experimental methods usually considered typical of science hindered the development of practical answers to engineering questions while failing to enhance theoretical understandings of the problems under investigation.

Similarly, during the same period, the founder of the engineering study of soils, Karl Terzaghi, had repeatedly stressed that no formulas for the engineering behaviour of soils *can possibly be obtained except by ignoring a considerable number of vital factors*.[698] The core equation in Figure 16.9 was first developed during the AASHO road test in the late 1950s (Chapter 20).

The Structural Number (SN) was the sum of the thicknesses of each pavement layer with each thickness modified empirically to account for its structural properties and way in which it was drained. The SN became an important part of many design, and assessment methods and techniques were developed to estimate it for existing pavements by *in situ* pavement testing – for example by using a device called the Falling Weight Deflectometer (Chapter 19). Such methods then made it possible to include the contribution of the subgrade to the structural performance of the pavement.

More recent design charts can be much simpler and often attempt to accommodate local conditions. A typical example is shown in Figure 16.10

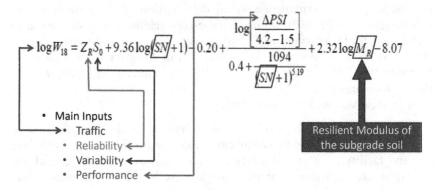

Figure 16.9 Pavement design chart based on 1993 AASHTO American national design code.[699] The equation is solved for the "structural number", SN.

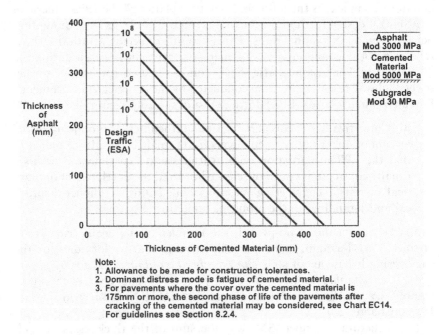

Figure 16.10 Modern simple pavement design chart.[700]

for a two-layer pavement on a given subgrade. Again, note the result typical of material fatigue findings in that an increase in required thickness is logarithmic and not linear. For example, increasing the traffic on a 200 mm thick bitumen-bound layer by a factor of 10 only increases the thickness of asphalt needed by a factor of about 1.3. This certainly helps to manage increased truck traffic.

The Strategic Highway Research Program (SHRP) was established by the American Congress in 1987 as a 5-year, $150-million research effort. Its objectives were to improve the performance, durability and safety of American roads. About a third of the research effort was spent on developing new specifications for bitumens and asphalt concrete mixtures. The work provided many new empirical links between key factors but did not deliver major new insights into the pavement performance. A subsequent second stage of SHRP concentrated more on safety, congestion, long-term pavement performance and the pavement renewal processes.

One output was the *Superpave* method for pavement design and construction. Its four key components were new bitumen specifications, new asphalt tests, a new design method and new maintenance procedures.[701] The Superpave mix design method was designed to replace the *Hveem* and *Marshall* methods (Chapter 15) by use of a gyratory compactor. Superpave design retains the earlier volumetric methods for mix design but also considers traffic and climate. Nevertheless, it somewhat unrealistically still lacked any test relating to material properties and it required relatively time-consuming, specialised and expensive testing equipment.

The availability of more powerful tools for computer-based stress analysis allowed a better understanding of pavement performance. This was broadly termed mechanistic design as it based on treating the pavement as a structural mechanism that could be rationally analysed. This was associated with an increased recognition of the role of each asphalt layer in the design. This was a significant change from the earlier notion that asphalt of lesser quality was required at the base of a pavement as the stresses there would be lower. Instead, it emphasised such issues as that the asphalt surface layer would need properties such as rut-resistance, whereas the bottom layer would need to have fatigue-resistance. This change went hand in hand with the shift from SN and the like to mechanistic-empirical methods of design. It required the stiffness properties of asphalt and such service factors as temperature, loading rate and loading frequency.

A further insight from this work provided was that the subgrade could not be regarded as a passive part of the pavement. When the pavement is loaded, the vertical stresses spread out, and the reduced stresses are dispersed through the depth of the pavement (Figure 16.3). Nevertheless, the integration of these reduced stresses over an extended area still has to equal the load the truck places on the upper surface of the pavement. They then cause the subgrade to deform and the extent of this deformation, governed by the subgrade properties, alters the distribution of stresses within the pavement.

A third factor was the recognition that existing tests did not adequately predict the compaction of asphalt under traffic, due both to the closure of voids and the breakdown of individual stones. This led to the under-prediction of wheelpath rutting. It was addressed by introducing a compaction test which included gyratory kneading of the cylindrical asphalt sample.[702]

Asphalt has increasingly dominated the pavement market, rising from some minor use in cities at the turn of the 20th century, to about 50% of total American pavements in 1970 and rising today to about 90%.[703] The asphalt (and concrete) pavements of the later 20th and early 21st centuries do not represent major structural changes and are well documented elsewhere[704] and so will not be pursued further in this study. However, one form which has survived for over a century for use on modestly trafficked roads is the use of a macadam-style basecourse of broken stones providing structural strength and covered by a surface course of asphalt about 100 mm thick.

It might appear to the reader that this chapter has sometimes been critical of the way in which asphalt paving technology has developed since about 1900 when asphalt entered its modern era. However, it must be emphasised that, unlike bridges and transport vehicles, pavements cannot be quarantined into neat and tidy analytical packages. Pavements are made from real materials, often sourced locally, and placed on ground with properties that can vary with time and place, and often in an unexpected way. The construction process is robust rather than precise, and the completed product is instantly buried out-of-sight. The constructor's priority will be to minimise the first cost, and any whole-of-life costing will be irrelevant. The users of the pavement are unpredictable in size and frequency and only loosely regulated. They view the pavement as a free facility managed and maintained by some invisible other. The pavement will degrade from the time it is first constructed and yet its maintenance and rehabilitation will often receive little priority. In these circumstances, it is little wonder that pavements are at a far different end of the engineering spectrum to bridges and vehicles. One must therefore admire rather than criticise those who have contributed to the development of pavements over the last hundred or so years.

Chapter 17

The 20th-century development of concrete pavements

Pavements for 20th-century traffic required some new emphases. As discussed in Chapter 13, pavements made of broken stones bound together by cementitious or bituminous mortars produce coherent structural slabs and can make excellent pavements meeting all practical operational requirements. However, there are major qualifications. Cement-bound pavements are very strong (rock-like) in compression but do not have any significant tensile or shear strength. Their low tensile strength means that they can crack under load, when expanding as temperatures rise, or when shrinking as the mortar dries. This cracking is managed, at a cost, by making the pavement relatively thick or by reinforcing it with steel bars. Indeed, in most markets concrete pavements have proved more expensive than asphaltic ones.

As discussed in Chapter 2, cement and cement mortars had been used by the earliest Mesopotamian builders and extensively by the Romans. The techniques for making cement were then largely lost until 1796, when James Parker ground and burnt lime-clay nodules he had found on the Isle of Sheppey in Kent and used the material to produce a practical cement. In 1824, Joseph Aspidin, a bricklayer from Wakefield in Yorkshire, invented modern cement by modifying the clay content and the burning temperature. His product was called Portland cement as it resembled in colour of limestone quarried on the Isle of Portland in Dorset. In 1825, Telford used the cement to make a 5.4 m wide and 150 mm thick concrete slab. This was employed for 2.5 km of pavement near Highgate Arch in London as a test for use on his newly commissioned Holyhead Rd[705] Broken stones were rolled into the top of the slab to provide a traffic surface.[706] The test was successful.

Concrete was used in the 1830s by Charles Penfold on relatively poor ground along the Brixton Road in South London. The concrete was mixed by hand in situ and was placed on the formation to a thickness of about 150 mm. It was then covered with another 150 mm of broken stone. Although the product was quite expensive, the pavement performed much better than its predecessor and the method was also used on other nearby South London roads.[707] The covering of broken stone meant that any inevitable cracking of the concrete would not be noticed and any resulting vertical movement of the cracked concrete would be remedied by the then-conventional topping up at the surface with further broken stones (a process then called *lifting* the road and discussed in Chapter 9).

Another form of concrete pavement was built in 1846 when concrete was used to carry a surface of granite setts on the Strand in London (see setts being placed in Figure 9.7). In the absence of steel reinforcing, the concrete cracked readily. In 1865, Joseph Mitchell, who had trained with Telford, introduced the cement penetration method (Chapter 13) in which a cement-sand mortar spread on top of a macadam pavement layer was swept and pushed into the interstices in the macadam course, aiming for the mortar to penetrate through to the bottom of the course. This was sometimes done in a number of layers and/or aided by rolling the surface to ensure full penetration.[708] The process was helped by using larger stones on the lower courses which led to much lower construction costs. However, of Mitchell's three trials, two – at Inverness and St James Park, London – failed due to early cracking. His third pavement, placed on the George IV bridge in Edinburgh in 1866, lasted for 40 years. The improvement was achieved by rolling a layer of broken stones into the top of the previous final surface. This approach came to be widely used.[709]

Rather than use concrete as a full pavement, as discussed in Chapters 10–15, in the 19th century concrete came to be widely used as a basecourse, typically 150 mm thick, between the firm subgrade and the surface course of asphalt or some form of segmental blocks. The concrete quality was relatively poor, and the thickness was chosen based on the past success-ful experience. The idea of using concrete to simultaneously act as basec-ourse and surface course was put to one side. This was possibly because the machinery did not then exist to allow concrete to be produced and then placed in the quantities and to the finish needed. Indeed, the concrete had often set before it could be brought to its final shape and texture. This led one perceptive commentator of the time to note that the concrete basec-ourse was the major contributor to the success of most asphalt pavements, which he noted would be better described as concrete pavements with an asphalt surfacing.[710]

From 1875, the use of improved construction methods saw an expanding use of concrete pavement slabs. Mixing cement, sand and stones to make concrete and then placing it in large enough quantities for road-making was not a simple task. The concrete was mixed by workmen on roadside platforms and carted to the pavement in wheelbarrows. The process used in Liverpool at the beginning of the 20th century is shown in Figure 17.1. The concrete was then compacted into place with hand rammers (Figure 9.7). The whole process was slow and difficult to organise and control. One solution was to place the course of broken stones first and use the penetra-tion method.

Liverpool System of "Hand" Concrete Mixing.

Figure 17.1 Manual mixing of concrete in Liverpool at the beginning of the 20th century.[714]

In 1892, the square outside the Main St courthouse in Bellefontaine in Ohio was paved by placing concrete directly on the subgrade. Its purpose was to prevent horses tethered outside the courthouse from destroying the existing macadam pavement.[711] Parts of this pavement were still in service in 1967.[712] However, many municipal authorities objected to the use of concrete slabs as they made access to services difficult.

Concrete would have been elevated in engineers' perceptions when, in 1899, Francois Hennebique built a triple arch (40 m, 52 m and 40 m spans) concrete bridge over la Vienne River in Châtellerault in France.[713] He reinforced the concrete with steel bars in order to control cracking in the otherwise brittle concrete.

The widespread availability of powerful construction machinery early in the 20th century increased the use of concrete slab pavements. Nevertheless, road-makers still often covered a typically 150 mm thick concrete slab with a surface layer of up to 100 mm of asphalt. Five reasons have been advanced for this.[715]

1. The surface of the early concrete slabs was too slippery for horse traffic.
2. Operators of horse-drawn vehicles preferred asphalt surfaces.
3. The concrete was rapidly abraded by horses' hooves.

4. Many of the early concrete slabs cracked badly in tension and were of low compressive strength. The cracking was magnified by the fact that the slabs were built without joints.

5. The concrete could take some weeks to attain its full strength.

At the other end of the spectrum, once concrete achieved full strength, it was very difficult to construct trenches to carry utility services below the pavement surface and to subsequently repair those services.

It was not initially appreciated that the strength of a cement mortar decreased as water was added to it, although adding water made the mortar easier to place. On the other hand, once the mortar had set it did not attain full strength unless its surfaces were kept moist. Both these factors ran contrary to the needs of low-cost construction and often led to sub-optimal outcomes.

Reports of such issues were magnified and promoted by the competitive asphalt industry and were only slowly resolved. Writing in 1903, a leading American road expert, Prof. Ira Baker, commented:[716]

> Concrete is much used for the foundations of pavements, and in a few cases it has been used as a wearing course. This form of road surface is not likely to come into general use owing to its cost and slipperiness ... and lack of durability.

On the other hand, quality control with concrete is relatively simple and was quickly codified. The proportions and characteristics of the components being used can be readily measured and observed. Test cylinders of the resulting concrete are taken from each mix and then tested in compression (crushing) after 28 days, the time in which the concrete is expected to be close to its full strength. The test load is applied to the flat ends of the cylinder. The testing process is universal and widely accepted. The test was first used in Germany in 1836. Thus, the quality of concrete is relatively easily controlled by the quality of the manufactured cement, by restricting the water used during mixing and by compression tests of concrete cylinders cast from the concrete being used in the pavement.

Technically, at the beginning of the 20th century, there was still no experience with the use of concrete as an independent pavement and methods for estimating the stresses in a concrete slab on a less than rigid subgrade were still decades away. The experience problem was addressed in 1909 when a kilometre or so of concrete pavement was constructed on Woodward Avenue running north from Detroit in Wayne County to the Michigan State Fair grounds. About a kilometre of the Avenue was paved between Highland Park and Palmer Park. Expansion joints were provided every 8 m.[717]

A key reason for the focus on Wayne County was that road-making in the region was not easy as it did not have good rock for making broken stone and had to rely on gravel and soft limestone.[718] The chosen design was validated by full scale tests conducted by the Michigan Public Works Department and discussed in detail in Chapter 20. The first major concrete highway constructed in the United States was a 40 km pavement built near Pine Bluff, Arkansas, in 1913.

After 1916, concrete roads were being built 100–200 mm thick, but little was known about thickness requirements. By 1919, favourable American experience with concrete roads was being noted in other countries.[719] This experience was greatly aided by the development of better mechanical methods for mixing, delivering, compacting and surface-finishing of concrete which permitted the efficient placements of pavements at least 150 mm thick.

Cracking remained an issue. In the early 1920s, many concrete pavements about 5 m wide were built in the Wayne County/Detroit area with no joints and later with a thickened centre section in a largely unsuccessful attempt to prevent the erratic longitudinal cracks which developed in most of the initial pavements. The cracking occurs when concrete is placed in tension at the base of the slab under traffic loads or throughout the slab due to concrete shrinkage being restrained by other parts of the pavement. The early experience showed that concrete pavements were particularly prone to corner and edge cracking if the subgrade support was inadequate. It was commonly found that it was cheaper to improve the support, rather than to thicken the slab at the corners and edges.

The isolated field trials had been unsuccessful and as Illinois was experiencing strong demands for new and better roads, to further advance the design and construction of concrete roads between 1912 and 1923, it conducted the Bates Test Road (discussed further in Chapter 20). The results from the Test Road led to the use a longitudinal centre joint to eliminate longitudinal cracking.[720] It also helped to develop simple analytical methods to predict pavement stresses, strains and deflections.

The amount of cracking in a concrete pavement depends on the traffic loads, the slab thickness, the stiffness of the subgrade and temperature changes. Thermal cracking due to temperature change occurs when the colder pavement surface is prevented from contracting by warmer lower layers. Shrinkage cracking occurs when concrete dries and hardens over time. The use of joints (i.e. cuts through the thickness of the slab) lets much of the cracking occur at manageable locations. Construction practice initially emphasised a full-thickness longitudinal joint on the centreline of the pavement slab. Transverse joints about every 15 m were then introduced, although subsequent experience saw these replaced by joints at about a 5 m spacing to manage thermal and shrinkage cracking. If reinforcing is used but is not continuous, a pavement may still need transverse joints about every 25 m.

An alternative to jointing was to follow structural engineering practice and add reinforcing steel to the concrete using rods or meshes to reduce concrete cracking. At the 1923 PIARC international road congress in Seville, there was still strong debate as to whether it was better to build concrete pavements as unreinforced mass concrete or as thinner reinforced concrete.[721] By 1935, jointing and reinforcing provisions for concrete pavements had been greatly clarified.[722] Major concrete pavements were built in Germany in the 1930s, including lengths of the State Highways from Munich to Starnberg and Munich to Augsburg, to examine various methods for using steel reinforcement. Prestressing pavements with tensioned cables to eliminate tensile stresses in the concrete were a European innovation following World War II.[723] Continuous reinforcing was introduced in the 1960s to avoid any need for joints. A 1970s innovation was fibre-reinforcing in which the use of steel reinforcing is avoided by adding small fibres of steel to the concrete during mixing. As discussed in Chapter 14, the fibres give the concrete an inherent tensile strength and so not only enhance the flexural strength of the concrete but also reduce deleterious superficial cracking that can occur above a length of reinforcing. Local economics then determines which of the four methods – jointing, jointing and local reinforcing, continuous reinforcing, prestressing or fibre reinforcing – is cheapest.

If reinforced, concrete pavements are a relatively straightforward exercise in structural engineering, compared with asphalt pavements (Chapter 17) which are far less elastic, and weaker in compression, particularly in high temperatures, and so are much more prone to rutting under traffic. Back in the 1920s, Clifford Older, the Illinois State's Highway Engineer, had produced some equations giving a concrete thickness inversely related to the tensile strength of the unreinforced concrete.[724] This was a tad optimistic as that strength would usually be close to zero and the measured tensile strength was probably better seen as an indicator of concrete quality.

Methods for predicting the stresses in idealised uniform elastic slabs loaded on their upper surface had been around since 1885 when Boussinesq provided a solution for stresses below the surface of a layer of elastic material due to a point load on its surface.[725] The practical breakthrough with concrete pavements did not occur until 1926 when Harold Westergaard published equations for determining the stresses in and deflections of loaded concrete pavements.[726] His solution assumed a concrete pavement to be a thin elastic plate resting on an elastic subgrade. This allowed the interior, edge and corner stresses in the plate to be calculated. To account for the stiffness of the subgrade, he introduced a term called the *modulus of subgrade reaction* which estimates how much the subgrade surface will deform vertically for a given pressure applied to its surface.[727] Westergaard himself only saw it as an empirical measure but others developed methods of assessing it by measuring the deformation of the subgrade surface when loaded by a circular plate (typically of 750 mm diameter).

Tests in the early 1930s by the American Bureau of Public Roads in Arlington, Virginia (Chapter 20) calibrated and modified the original Westergaard equations, and for many subsequent years, they provided the basis for the design of concrete pavements. Generally, they suggest that the thickness needed increases with the square root of the wheel load and that an unreinforced pavement has to be about 40% thicker than a reinforced one. However, it remained common practice to assume that the pavement would be 150 mm thick, unless special circumstances suggested otherwise.[728]

In 1943, Donald Burmister developed a solution for the stresses in a two-layer pavement and in 1945 for a three-layer pavement.[729] The design process was modified in 1946 by the American Portland Cement Association (PCA) to include new data from cyclic load (fatigue) tests and, in the 1960s, methods of analysis allowed multilayer pavements to be designed.[730] In 1951, design charts were produced to handle the real-world case where loads come from multiple tyres. More recently, reliance has been placed on *finite element* methods of stress and strain analysis. A major step forward occurred in 1984 when the PCA issued a new design manual for concrete pavements in which pavement thickness is selected to minimise failure of the pavement by fatigue cracking due to the passage of wheel loads.[731] The link between the number of wheel passes and pavement thickness is similar to that used for asphalt pavements (Chapter 16).

One problem that arose with early concrete pavements was that if water was trapped between the pavement and the subgrade, the pressures caused by traffic travelling along the pavement would pump the water and associated subgrade materials out of the subgrade via joints, cracks and pavement edges (Chapter 1). This could leave the pavement inadequately supported and prone to structural failure and required increased attention to subgrade design and drainage. One solution was to place a layer of open-graded broken stones between the concrete pavement and the subgrade to act as a drainage layer. This was normal practice until the 1980s, but it is now more common to use some form of bituminous- or cement-bound material under highway pavements. This non-structural layer also serves three other purposes. It rectifies soft spots in the subgrade which can affect stiff pavements, accommodates relative movement at joints, and minimises any shrinking and swelling affects (Chapter 1) in the subgrade.

A concrete pavement is inherently less flexible than an asphalt one, and so greater emphasis must be placed on the stiffness and uniformity of its subgrade to ensure that it does not gradually settle vertically and thus fail to adequately support the pavement. Recall that in Chapter 1, it was sufficient to define an acceptable subgrade as "firm" and that Chapter 15 then showed that subgrade stiffness became even more important as methods for the stress analysis of the pavement structure improved. If the basecourse was too thin, its cracking could result in erosion of parts of the subgrade. Consequently, as traffic loads increased in frequency and size, subgrade performance became even more relevant. Recall that one onsite method of subgrade assessment had been to take a heavy roller over it and note any deformation – particularly if it persisted after the roller had passed. The next step was to measure the deformation in a more standardised way via the modulus of subgrade reaction described above.

Concrete pavement construction boomed during the 1930s. For instance, in 1926, there were 300 km of concrete pavements in Britain. The many improvements in concrete pavement technology discussed above meant that, by 1935, the length was 5,000 km. Germany began building its autobahn system in 1933, after Hitler came to power. By the end of World War II, it had constructed some 4,000 km of road. They were almost entirely of concrete placed on "mass production lines"[732] but in a manner which suited local conditions and the desire for the work to be performed by otherwise unemployed and often unmotivated workers. Steel for reinforcement was often unavailable and many of the pavements used plain concrete. The pavements were typically 250 mm thick and jointed every 10 m. The lower layer was 150 mm thick and used locally available stone. The upper 100 mm layer used high-quality stone. The subgrades were well drained.

The zenith of concrete pavement usage was possibly reached in 1938–1940 when America's first paved tollway, the 270 km Pennsylvania Turnpike, was constructed using concrete for all of its pavements. In retrospect, it is not clear how a concrete pavement rated so highly in the Turnpike engineers' perspectives, but the selection is consistent with the decision to use concrete for the German autobahns. A major German commentator has suggested that German engineers had been impressed by American concrete pavements, as reflected in their road magazine Verkehrstechnik between 1926 and 1930.[733] However, when the second American tollway, the 80 km Maine Turnpike, was built in 1947, it only used asphalt paving. The same outcome in favour of asphalt occurred with the massive New Jersey Turnpike built in the 1950s.[734] From that time forward, asphalt has dominated many paving markets.

In what was possibly a clever marketing ploy, over time concrete pavements were often called *rigid pavements* and asphaltic pavements were known as *flexible pavements*. Technically, all pavements are flexible and no pavements are rigid. The degree of flexibility of a pavement depends on the stiffness of pavement layers and the subgrade. Stiffness – load/deformation – depends on pavement material properties and on the thickness of the pavement layer.

The commencement of US Interstate Highway system in the 1950s saw two major influences that have shaped world practice in concrete highway pavements. The extent of work allowed major developments in large volume concrete production plants and slipform paving. This made unreinforced concrete with short slabs simpler and cheaper both to construct and repair. Usage of the older longer slab reinforced pavements declined. Since around 1970, the great majority of international highway construction has been in unreinforced 'plain' concrete as have been airfield pavements. However, significant lengths of continuously reinforced pavements have been constructed in Texas, Illinois and Belgium.

Two of the major advances to occur in recent times were, first, the introduction of vibrators to aid the placement and compaction of concrete which had a major positive effect on concrete quality. The second advance was the introduction of slipform paving. In this process a single, self-powered machine converts wet concrete into a finished pavement (Figure 17.2). Travelling at a walking speed on caterpillar tracks, it is able to place concrete which is sufficiently stiff to maintain the pavement profile without the need for side formwork. Tensioned string-lines or laser guidance is used to ensure the correct longitudinal placement of the concrete and a mechanical vibrating screed provides the required surface finish. The steel reinforcing is placed by the machine.

A pavement surface must provide adequate friction for braking and turning vehicles (Chapters 2 and 19). To achieve this with concrete pavements, the new surface must be broomed, grooved or patterned or the mortar removed from the surface to expose the tops of the stones used in the concrete mortar. Typically, a concrete pavement will need its surface texture to be restored every 20 years.

Roller Compacted Concrete (RCC) is a non-reinforced concrete pavement. It was first used in the 1970s when the Canadian logging industry switched to land-based log-sorting methods which required low-cost, strong pavements able to carry the large wheel loads encountered in the sorting yards. RCC has the same basic ingredients as conventional concrete: broken stone, sand, Portland cement, supplementary cementing materials, chemical admixtures and water. The mix uses a similar aggregate gradation to asphaltic concrete (Chapter 15) and so can be placed directly from agitators onto the subgrade using similar paving equipment and vibratory rollers. The mix is stiffer than conventional concrete and so can be placed without formwork or surface finishing. As reinforcing steel is not used, any unwanted cracking is managed by the thickness of the concrete and joint spacing. One difference between RCC and conventional concrete is that the

Figure 17.2 Modern slipform concrete paver.[735]

texture of the final surface is more open and similar to an asphalt surface and can lose some fine aggregate in the initial years of service. This loss can be minimized by grinding the surface after construction. If used in higher-speed applications, RCC usually requires a surface layer of asphalt or grinding of the concrete to ensure an adequate ride quality.[736]

The repair of concrete pavements has recently been advanced by the use of precast post-tensioned concrete slabs since about 2000. It reduces delay due to pavement repairs and increases the service life of the subsequent pavement.

Chapter 18

Chemical and physical means of modifying local materials for paving

Because of the nature and location of pavements, it is often not feasible to use the best possible materials in their construction a pavement. Commercial and technical compromises must often be made. One approach is to modify the sub-optimal materials to enhance their appropriate properties. At the other end of the spectrum, pavement and subgrade soils can be enhanced by processes broadly called *soil stabilisation*. The use of the term *stabilisation* is curious as the role of the associated processes is to enhance, rather than to stabilise. The use of bitumen and cement to bind stones together, as discussed in Chapters 13 and 17, lies at one end of the enhancement spectrum. Table 18.1 summarises the six common stabilisation methods.

Lime and cement stabilisation are remarkable but quite different processes to mortar-making in that the addition of a very small percentage of quick lime (calcium oxide), slaked lime (calcium hydroxide) or Portland cement (Chapter 11) to some soils can have a major overall impact. This is particularly useful with clays which are inherently unstable and very moisture sensitive.

The remarkable improvements in clay properties stem from the fact that particles of these materials are surface reactive, with many free surface ions which preferentially link to clay particles. This binds the particles and prevents water ions from causing the adverse effects, as discussed in Chapter 4. A key consequence is that the amount of cement or lime required is much smaller than would be used in making a normal cement mortar. The process reduces the moisture susceptibility and increases shear strength (and thus the bearing strength) of the material. However, as it works at a particle level, it is only effective if the additive can be thoroughly mixed into a pulverised soil.

In cement stabilisation (Category 3 in Table 18.1), the amount of cement used is sufficient to turn the material into a structural layer. However, it can still exhibit a pattern of widely spaced cracks.

The stabilisation technique may have been used by ancient road builders but modern applications only began in the American south in 1906, followed by the granting of American patents in 1910. A further patent for a product called *Soilamies* for highway use was issued in 1917. The first field

Table 18.1 Stabilisation categories and characteristics[739]

Category	Process adopted	Anticipated performance attributes
1. Strongly bound cemented materials	Addition of greater quantities of cementitious binder, commonly > 3% (see Chapters 13, 15 & 17)	Increased pavement stiffness May be susceptible to fatigue cracking May introduce transverse shrinkage cracking
2. Granular stabilisation	Blending with other granular materials, as in Chapters 4 & 9	Improved pavement stiffness, shear strength and resistance to aggregate breakdown
3. Stabilised subgrade materials	Lime and/or cementitious binder added to high plasticity soils such as clays and fine silts Cement and/or cementitious binder added to low plasticity soils such as coarse silt and sand Bitumen and foamed bitumen	Improved constructability, subgrade CBR, stiffness and shear strength Reduced heave and shrinkage
4. Lightly bound cemented pavement materials	Addition of small quantities of cementitious, lime or chemical binders commonly < 3%	Improved pavement layer stiffness and rut-resistance Reduced sensitivity to loss of strength due to increased moisture content But may be susceptible to: fatigue cracking occasional premature distress at low binder contents, may be subject to erosion where cracking is present
5. Naturally cementitious materials	Slags, fly ash, pozzolans	Similar to Category 4
6. Mechanical stabilisation	Grading of stone sizes, rolling and compacting (see Chapter 4)	Improved pavement strength, stiffness and water resistance

use of cement stabilisation appears to have been in South Carolina in 1932, and the method was formally codified by the American Portland Cement Association in 1935.[737] It was used with great success to build airfields during World War II.[738]

In situ stabilisation is a mobile process in which a purpose-designed machine is used to break up the existing pavement, add the stabilisation material (typically cement or lime) to the broken pavement, mix the materials together and then compact and level the mixture. A typical modern machine that does these tasks in a well-controlled manner is shown in Figure 18.1.

Figure 18.1 Modern Wirtgen soil stabilisation machine.[741]

The quality of the resulting *in situ* stabilised material is often variable, and in some cases, it is necessary to add other unbound granular materials to correct deficiencies in particle size distribution and/or plasticity. The in situ stabilised material is then cured prior to the placement of overlying layers such as additional stabilised or unbound granular layers or bituminous surfacing.[740] Typically, the amount of cement used is about 4% of the material being treated. Modern machinery allows depths of 400 mm to be treated, but this requires some additives in the cement to prevent the mix setting before the treated layer has been compacted and levelled.

Some manufactured materials have inherent stabilising properties (Category 5) and can be excellent pozzolans. This particularly applies to the *crushed slags* (or *ground granulated blast furnace slags*) produced as consequence of iron- and steel-making. It acts as a slow-setting cement by itself and also reacts well with lime and so it is an excellent pozzolanic material and is used in pavement stabilisation. Pavements constructed from these slags can become fully bound through natural cementation. However, this is particular to the slag source and is not a general property. Some crushed slags already contain small amounts of free lime but to ensure cementation occurs they may need to be pre-blended with lime.[742] Crushed slags used as granular pavement layers generally conform to the conventional standard specifications (particle distribution, plasticity and hardness) of road authorities.

Fly ash is a product of the power-generation industry. The type of coal used, and the mode of operation of the plant, determines the chemical composition and particle size distribution. Consequently, not all sources of fly ash are suitable for stabilisation. Generally, fly ash derived from the burning of black coal is high in silica and alumina and low in calcium and carbon, leading to it often having pozzolanic properties. Hence, it is usually well suited for use in stabilisation. On the other hand, fly ash derived from the burning of brown coal contains large percentages of calcium and magnesium sulphate and chlorides and other soluble salts, and hence, it is usually unsuitable for use in stabilisation.

Stabilisation of soils with bitumen or other similar polymers (Category 3c) is akin to the use of cement to form a soil-based mortar (Category 1). The bitumen coats the soil particles and binds them together as a mortar. To make it possible to mix bitumen into a soil, the fluidity of the bitumen is increased by adding fuel oils and other fluid hydrocarbons, by emulsifying it with water, wax or soap, or by foaming it with steam. It is not very effective with silts.

Foamed bitumen remains viscoelastic and is resistant to high in-service temperatures. It is produced by introducing small amounts of water (2%–3%) to hot bitumen. The water evaporates instantaneously, forming gaseous bubbles which cause in the bitumen to foam and expand by 10–30 times its original volume. This increased volume results in a reduction of the viscosity of the hot bitumen, allowing it to mix with a wide range of materials, including ones which are of small size, cold or damp. It also has an enhanced ability to wet and then adhere to any stones in the mix. In particular, the bitumen bubble film encapsulates any small particles in the material being stabilised. The bubbles burst in a few minutes, restoring the properties of the original bitumen. On the other hand, the alternative emulsified bitumens can take hours to "break" and restore bitumen properties.

Mechanical stabilisation (Category 6 in Table 18.1) is a term that stems from the fact that a problem with building pavements in areas without suitable stone but with ample sand, soil and clay was that it was often difficult to compact such materials into a suitable dense pavement structure. A pavement layer can prove unstable if it is not compacted. In fact, compaction produces three main beneficial outcomes. First, it minimises any future settlement of the layer under traffic. Second, it helps the traffic loads to be carried through the layer by increasing direct stone to stone contact, which is the strongest and stiffest load-carrying route. Third, it makes the pavement less permeable by minimising air voids and thus protects lower layers and subgrades from the effects of water.

Compaction requires energy to be applied by compaction equipment and there are limits to what can be achieved. Some water in a pavement layer aids compaction but excess water, due to the incompressibility of water, makes compaction harder. Furthermore, for a material such as clay, excessive compactive effort can break down the inter-particle structure and lower the strength of the pavement. Thus, as mentioned in Chapter 4, there is an optimum moisture content at which maximum compaction is achieved for a given level of compactive effort.

As discussed in Chapter 4, clays are dusty and brittle when dry but can quickly become soft and sticky when wet. Conventional rolling of such pavements (Chapter 13) to raise their density was often ineffective. Pneumatic-tyred rollers brought some better outcomes as they additionally tended to knead the pavement layer.

In California in about 1900, the sheepsfoot roller was discovered – almost by accident – and proved to be an excellent device for compacting clayey and cohesive soils.[743] The *accident* was the observation by two road builders, Walter Gillette and John Fitzgerald, of how well a flock of sheep which had strayed overnight across their unfinished oiled pavement (Chapter 14) had compacted that pavement. To replicate the effect, they hammered iron spikes into a log which soon morphed into the first sheepsfoot roller. A subsequent early version is shown in Figure 18.2. The device is now widely used. The *accident* was not unique as 18th-century canal builders had used herds of cattle to tamp into place the layers of clay used to waterproof the surfaces of their canals.

Dry clays, sand and most soils on their own do not make good permanent pavements as they are easily displaced by wind, water and passing traffic. Some communities with pavements made from these materials have achieved short-term success by spraying the natural road surface with waste tar or oil.[744] These measures particularly related to dust suppression, as discussed in Chapter 14. Loose gravels are marginally better than soil and sand as they are more resistant to the effects of wind and water. Mixtures of clay and gravel can be even more effective, particularly if the clay can be kept moist and thus act as a mortar between the pieces of stone. A version of this approach is the *water-bound macadam*, as discussed earlier in Chapter 9. Empirical measures and methods for material selection and placement to ensure good outcomes have been developed.[745]

Figure 18.2 Sheepsfoot roller. Albaret was a French manufacturer of compaction equipment.[746]

Figure 18.3 Road burner in Irvine, Queensland.[749]

Thermal stabilisation is a very different use of clay for making pavements. It was developed in parts of the New World where road builders found little useful stone but many trees and lots of clay. One method burnt a mixture of clay and coal in roadside ridges to produce an inert brick-like material.[747] A slightly more sophisticated version placed logs on top of a prepared road surface and then covered the logs with more clay. The logs were then burnt, and the remaining ash and burnt clay was rolled to produce the final pavement. The method was promoted by the American Bureau of Public Roads, and in the 1930s, it was extended by Queensland Main Roads to use mobile furnaces in which logs were burnt and the heat of the fire directed on to the clay pavement surface (Figure 18.3).[748] The lack of subsequent application suggests that the process was not cost-effective, probably because very high temperatures are needed to vitrify a clay to brick-like properties. However, it has recently been noted that heating to 300°C may be sufficient to destroy a clay's swelling and shrinking propensities.

Geotextiles are either coherent impermeable membranes or permeable fabrics in which threads are either woven, knitted or bonded to produce useful sheets. They are used for improving the crack-resistance of a pavement, protecting a subgrade from construction traffic, preventing mixing of two pavement layers, stopping water movement or acting as a drainage layer (Chapter 2). The materials used are commonly plastics such as polypropylene, although hessian and coconut fibres have also been employed.

Chapter 19

Devices and methods for measuring and evaluating pavement performance

So far we have discussed the various types of pavement, how they came about and how they were designed and constructed. We have also examined the mechanics of how they operate in situ and under traffic. It is now appropriate to observe how pavements perform in the real world and to understand how that performance is measured and assessed. This information is important for the provision of future pavements and the management of existing ones.

The prime purpose of a pavement is to provide travellers with a surface on which they can undertake their journeys to expected levels of ease, safety and operating cost. A well-constructed and maintained pavement should be able to meet these expectations. However, a pavement can fail to do so in four main ways.

1. It can provide an unacceptably rough ride for travellers and cargo by variable profile changes along its length.
2. It can make travel difficult when its surface fails locally by factors such as rutting, pot-holing, surface cracking, loosening of stones in the pavement surface (ravelling) or peeling free of the underlying pavement course (delaminating).
3. Its surface can be unacceptable when it is too slippery to meet the frictional needs of the wheels of modern vehicles, too glossy or covered with ponding water to meet the optical requirements of safe driving, too dusty for safe driving and acceptable roadside amenity, or too noisy for travellers and people adjacent to the road.
4. It can fail structurally and make travel very difficult and hazardous. It might be expected that structural failures will be due to traffic loads (Chapter 16). However, the preceding chapters have shown that local conditions, particularly those related to water and temperature, are also major factors in determining the structural performance of a pavement.

The short-term negative consequences of any or all of these four factors will be to increase pavement maintenance costs and vehicle operating costs and to reduce road usage. Because pavement deterioration raises user costs and accelerates further deterioration, the optimum pavement management strategy is usually to maintain a pavement in good condition.

Table 19.1 provides a summary of the performance targets needed for pavements to achieve positive outcomes in the light of these factors. The associated requirements can be specified in terms of actual numbers, but the nature of the natural materials used means that variability of the properties must also be understood. Pavements are built in the real world, and minimising costs will mean that they commonly use local materials. Real materials in the real world behave in complex ways and their properties vary greatly. Thus, while methods of material testing and structural analysis have become increasingly more sophisticated, there is still often a large gap between theory and practice. As a consequence, there remain many empirical and locally oriented factors in pavement design, and pavement managers often take some silent solace in the fact that the occasional repair of piece of failed pavement is often much cheaper than providing a long length of over-designed pavement.

The associated requirements for a completed pavement are to meet appropriate limits on the six factors listed in Table 19.2.

Not surprisingly, 19th-century engineers carried out research and conducted tests and experiments to try to understand what was needed to meet the requirements in Table 19.2, led by Irishman John Macneill and Arthur Morin in France (Chapter 13). Macneill's 1829 *road indicator* was the first device developed to specifically measure road condition. It used a dynamometer based on existing weighing machine technology using weights and lever arms to assess the force required to pull a vehicle at known speed along a road.[751] It was produced while Macneill was working with Telford on the Holyhead road (Chapter 9). However, the very rapid growth in

Table 19.1 Subgrade and pavement material requirements (based on data in[750])

Requirement	The product must be able to:	Test Criteria
Subgrade stability	Support pavement layers	Bearing capacity, resistance to deterioration over time
Workability in construction	Be placed to the required dimensions and conditions	Stone gradation, crushing strength, resistance to deterioration
Operating under traffic	Carry applied loads without unacceptable deformation, be stable in a changing local environment and be durable over time	Strength and stiffness of the compacted layer, resistance to breakdown and pot-holing under load and water penetration, minimal roughness, minimal ponding
Resistance to surface wear	Resist surface abrasion, erosion and attrition	Skid resistance, polishing (abrasion) resistance, temperature stability

Table 19.2 Pavement operational requirements directed at minimising the following six factors:

Factor to minimise	Description
Whole-of-life cost	Achieve an economic balance over the pavement's intended life between the cost of construction and the cost of maintenance. All pavements eventually require restorative maintenance, typically by placing an asphalt overlay on top of the existing pavement, although in urban areas the existing surface may require removal of old material by milling in order to preserve critical surface levels and crossfalls relative to adjoining infrastructure. On a toll road, the minimisation would also consider toll revenue
Transverse profile	Select materials and design and construction processes to minimise rutting caused by high vertical wheel loads, many repetitions of wheel loads, poor compaction, bitumen softening, shrinking and swelling of underlying materials and surface wear
Longitudinal profile	Manage construction to minimise non-uniform deformation, profile changes of the pavement along its length (roughness) and major compaction of a pavement layer under vertical wheel loads. Select materials to avoid shoving of a pavement surface under horizontal braking and traction loads
Slipperiness and glossiness	Select materials and bonding methods to minimise polishing or abrading of the surface under traffic or accumulating sheets of water during rain (Chapter 2)
Surface (or alligator or chicken-wire) cracking and ageing disintegration	Select durable bonding materials. Design and construct to prevent spalling at joints, thermal and fatigue cracking, upwards pressure from water in pavement pores, and secondary consequences of rutting. Select pavement surfacing materials to avoid ravelling, stripping and bleeding of bitumen
Gross deformation	Design and construct to prevent subgrade failures, pumping of subgrade materials, gross shrinkage and swelling, and water eroding supporting materials

motor traffic, in number, size and weight of vehicles in the late 19th century outstripped this approach.

Before the advent of the motor vehicle, the selection of construction materials centred on the properties of gravels and broken stones, principally in relation to particle size, crushing strength, impact resistance, durability, porosity and abrasion resistance (Chapter 4).[752] Roads have always required some sort of maintenance, and the 'life' of pavements was commonly thought of in years before any need for substantial reconstruction. However, the predominance of pavements built to 19th-century expectations but, even early in the 20th century, carrying pneumatic-tyred motor vehicles of greater weight, travelling at greater speed and in higher numbers, meant that it soon became apparent that there was a need for higher quality construction materials and a better understanding of the strength and stiffness of the subgrade and of the selection of pavement materials and thicknesses. In addition, inadequacies in the methods used for the stress analysis of pavements discussed earlier (Chapters 16 and 17) meant that flaws were often magnified.

Probably the major single issue in this new regime related to traffic loads. Given the expected environmental conditions, the prime purpose of a pavement is to carry traffic and so its key performance indicator is how well it does that task in the given environmental conditions and when subjected to the wheels and hooves of the passing traffic. The four key traffic task measures are the size and frequency of the imposed vertical and horizontal loads transmitted via the wheels and hooves.

Before the 20th century, a major traffic concern was that a solid wheel with iron "tyres" could exert very high local pressures on the pavement surface, particularly if that surface was uneven. An otherwise well-constructed pavement would rut in a matter of days, and the problem was often exacerbated by the ruts then collecting water which was able to percolate into the pavement. The ruts also concentrated the vertical loads in a very narrow wheelpath. Rut depth in an existing pavement could be simply, but slowly, measured by using a straight-edge.

It was not uncommon for cart and wagon wheel loads to be as high as 2 tonne, and loads as high as 1.5 tonne were acceptable to most pavement administrators.[753] A single overloaded wheel could totally destroy a length of pavement, and so administrators were alert to the prospect. However, initially their only tools for detecting overloads were visual observations of the vehicle and its cargo and of the effect of a wheel on the pavement immediately beneath it. In practice, wheels loads were mainly limited by the design of the vehicle and by the ability of the haulage animals to pull the loaded vehicle (Figure 19.1). The first device for weighing carts and wagons

Figure 19.1 Load-limiting process in early Victoria, Australia.[756]

was a lever arm device introduced in Dublin in 1555 – such *weighbridges* were used more to collect tolls based on the load carried than to prevent pavement damage. Pioneering work in France[754] in the 1840s showed that the stresses that a wheel load produced in a pavement were inversely proportional to the diameter of the wheel. The effect was mainly a consequence of an increase in the contact area between the wheel and the pavement.[755] This understanding led to some very large wheels which created a new set of problems as they were heavy, expensive, raised the cargo even further off the ground, and made steerable axles impossible as wheels on a turning axle could not pass under the body of the vehicle.

The situation became less critical over the 20th century with the widespread use of pneumatic tyres and of effective loading controls. The solid rubber tyres which had prevailed from about 1890 were marginally less damaging than iron-tyred wheels and produced about twice the contact pressure as did modern pneumatic tyres. Pneumatic tyre sales surpassed solid rubber tyres in the mid-1920s. A pavement advantage of pneumatic tyres is that they cannot exert a pressure on the pavement greater than the tyre pressure. Most modern (non-military) wheels impose maximum vertical loads between about 1 and 2 ton. An associated administrative advantage of pneumatic tyres is that if overloaded they will show it very obviously by bulging and quickly bursting. This is an important attribute as a single pass of an overloaded wheel can cause major pavement damage. Very heavy cargoes are accommodated by adding more wheels rather than by increasing individual wheel loads. The separation of the wheels ensures that their combined stresses are always less than each individual peak stress.

The loading issue is thus more the arrangement of the wheels that use the pavement, rather than the load that each can transmit. This directs attention away from overload and towards the fatigue failure of the pavement. This factor was highlighted towards the end of World War I when new American trucks destined for the European battlefields were driven over the 19th-century pavements between the inland American truck factories and the ports on the Atlantic seaboard. The resulting number of wheel loads and their impact on the existing pavements caused a well-publicised *virtual collapse* of the highway system in the eastern States.[757] A major recovery and reconstruction effort began in the post-War years, led by Thomas MacDonald who had been appointed Director of the Bureau of Public Roads in 1919, replacing Logan Page who had died late in 1918. MacDonald held the position for the next 34 years. His initial task at the Bureau had been aided by the desire of General John Pershing to demonstrate the future military effectiveness of trucks. He did this in 1919 by sending a fleet of 79 trucks and about 300 soldiers (including a Lt Col. Dwight Eisenhower) across America from Washington to San Francisco. The trip took 56 days and highlighted the need for good pavements as much as the value of trucks.[758] After he became President, Eisenhower later began the construction of the American Interstate freeway system.

Universally, the growth of motorised traffic on highway networks has continued to exceed any prior expectations. Although individual wheel loads remained reasonably well controlled, the number of wheels (technically, in terms of the number of equivalent standard axles, Chapter 16) has rapidly increased, placing greater emphasis on pavements failing by fatigue rather than by rutting. This most noticeably occurs at the pavement surface and is often addressed by applying a new layer of asphalt on top of the existing sound pavement. Fatigue failures are practically inevitable and even good pavements might require such a maintenance treatment every decade.

The way in which wheel loads are dispersed within a pavement is shown in Figure 16.3, and resulting strains are shown in Figure 16.4.[759] A worked example of the stresses, strains, deformations and deflections within a pavement is given in Figure 19.2. The ability of soils and similar materials to carry various stress combinations had been well developed by workers in the field of soil mechanics and those methods, such as triaxial testing of soil specimens (Chapter 4), can be readily transferred to the study of pavement performance so their history is not repeated here.[760] The calculation of these stresses and strains has three major complications.

1. Almost all pavements are multi-layered and so it is necessary to describe the degree to which they act in unison – technically, the degree to which horizontal shear strains can be transferred from one layer to another.
2. Pavement materials are rarely truly elastic – they deform more at higher stresses and they may deform slowly even when the stress levels do not alter (called *viscoelasticity*).
3. Tensile stress can lead to cracking under a single load or after many load repetitions (*fatigue*, Chapter 4). Note from Figure 16.4 that tensile stresses can occur at the pavement surface and can be a major cause of wheelpath pavement cracking. Nevertheless, fatigue provisions were unhelpfully hidden away in the design method, as shown in Figure 16.9.

To avoid using materials which subsequently prove to be not fit for their intended purpose, in recent times most organisations responsible for pavements have developed relatively detailed and comprehensive specifications for pavement materials. These requirements and tests are now effectively used worldwide in their application due to the work of international road associations such as PIARC (Permanent International Association of Road Congresses) and IRF (International Road Federation).

Specific tests have been highlighted in the preceding chapters. For example, as improved surfacings and better tyres permitted pavements to became thinner in the 1920s, Chapter 13 showed how the CBR test was developed to particularly relate pavement structural performance under traffic to the deformability of a subgrade. The full-scale testing of pavements under real

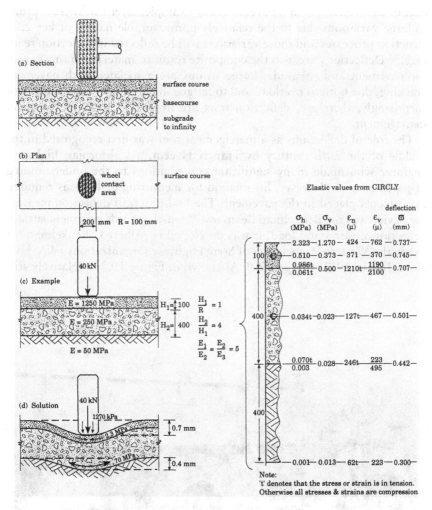

Figure 19.2 Worked example of the stresses, strains, deformations and deflections in a loaded pavement.[761] CIRCLY is the method of stress and strain analysis discussed in Chapter 16. MPa (megapascal) is the international unit of stress.

loads is described in Chapter 20. These tests are important as they give an opportunity to closely observe how real pavements perform under real traffic.

One obvious indicator of the structural characteristics of an existing pavement is how much it deflects (or deforms) vertically under a passing vertical traffic-induced load. The size of the deflection will depend on the thickness of the pavement and on the stress–strain response of its constituent materials and of the subgrade. It is thus a good indicator of the overall characteristics of a pavement. For example, layer thickness is prone to large variations due to the relatively immeasurable nature of key construction processes and those variations will be reflected in deflection readings.[762] Deflections are also the composite result of material strains within the pavement and subgrade. Large strains are associated with pavement cracking, due both to overload and to fatigue under repeated loadings. Not surprisingly, therefore, deflection tests are important tools in pavement management.

The role of deflections as a management tool was first recognised in the middle of the 20th century by Francis Hveem, a Californian highways engineer who made many significant contributions to the understanding of pavements.[763] However, his method for measuring deflections required instruments placed in the pavement. The first practical deflection measuring device was the Benkelman Beam test[764] which records pavement surface deflection under static load. It was developed by Albert C Benkelman for use in the Western Association of State Highway Organizations (WASHO) Road Test in 1952 (Chapter 20). As shown in Figure 19.3, a relatively stiff

Figure 19.3 Benkelman Beam testing during the WASHO Road Test (Chapter 20).[766] ACD is the supporting reference beam sitting on the undeflected pavement at A and C. BDF is the deflection beam, supported from the reference beam at C and resting on the loaded pavement at F. As it pivots around C due to the dual wheel loads at E, its vertical movement relative to ACD is measured by a deflection gauge at B.

beam is placed between the wheels of a stationary loaded truck. Observers then measure the pavements vertical deflections from the beam, assuming that the ends of the beam sit on pavement unaffected by the load. The plot of the deflections along the beam is called a *deflection bowl*. Note that a sound pavement would typically deflect less than a millimetre under a legal load, and so any measurements need to be taken with care.

The data reflect pavement stiffness and was often used in conjunction with the roughness measures to be discussed below. Despite its empirical nature, the test is still widely used, particularly a more mechanised French version known as the Lacroix Deflectograph first introduced in the 1950s. British field studies allowed the measured deflections to be empirically related to the remaining life of the pavement.[765]

To this stage, we have implied that wheel loads are applied slowly without any impact. Of course, this is far from the case, particularly with trucks speeding along rough roads, and so a number of non-destructive field tests were developed to measure pavement deflections under a range of dynamic loads. Such measurements of pavement condition play two key roles. First, they allow future maintenance strategies to be planned. Second, if restorative overlays are required, they can be designed on the basis of known, real conditions.

In recent times, devices have been developed to measure pavement deflections due to a truck travelling at highway speeds. For example, the Danish Traffic Speed Deflectometer vehicle uses Doppler laser sensors to measure the vertical velocity of the pavement surface as it is deformed downwards due to the vehicle wheel loads. The vertical movement is calculated by integrating the vertical velocities.

A related development was to simulate real transient traffic loads by applying an oscillating dynamic load to the pavement and measuring the resulting dynamic deflections. This was done using counter-rotating weights applied to the pavement and recording the associated surface with seismic accelerometers. A common device is the Texan Dynaflect machine. The loads applied are similar to actual wheel loads and so would not be expected to cause any observable pavement failures.

A similar and widely used device is the *Falling Weight Deflectometer* (FWD) which drops a weight onto the road surface and uses accelerometers on the pavement surface to measure the deformational response of the pavement and subgrade (Figure 19.4).[767] It was developed in Scandinavia in the late 1960s. Current versions can operate at a highway speed. The measured surface deflection of the loaded pavement is a common surrogate performance measure. However, the FWD software is also used to estimate the strength and stiffness of both pavements and subgrades at various load levels.[768] These data are useful for planning pavement rehabilitation and for checking new pavements.

A basic geometric property of a pavement is its longitudinal vertical profile. The core inherent component of the profile is the vertical alignment

Figure 19.4 A Dynatest FWD.[769]

of the road itself. The common international characterisation of main deviations from the core alignment is given in Table 19.3. The major transverse geometric concern is rutting.

The wavelength of the macrotexture relates the size of the individual stones used to form the pavement surface and so is usually less than 50 mm. Macrotexture is a critical contributor to pavement performance as it provides the essential resistance to vehicles skidding when braking or turning and minimises disturbing light reflectance. In the pre-automobile days macrotexture would also have related to road users, both two-legged and four-legged ones, having adequate foothold. It would have been simple to observe if such was not the case. Of course, pavement friction can also

Table 19.3 Common characterisation of the components of variations to a pavement's inherent longitudinal vertical profile[770]

Characteristic	Wavelength	Definition
Microtexture	< 0.5 mm	Asperities on the surface of stones in the pavement surface. They are worn away by traffic but often replaced by seasonal weathering. They influence pavement surface friction in dry weather
Macrotexture	0.5–50 mm	Relates to the frictional properties of the pavement
Megatexture	50–500 mm	Relates to major surface defects in the pavement such as potholes
Roughness	> 500 mm	Discussed extensively in the following pages

negatively influence the haulage task, and we saw earlier how Morin and Telford had arranged tests to quantify this relationship (Chapter 13). With modern high-speed traffic, the macrotexture of pavement surface plays an even greater role.

The frictional performance of a pavement surface is provided by both the smoothness of the individual pieces of stone and by the texture of the surface. The first effect is the result of resistance of molecular bonds between the stone surface and the tyre rubber and dominates in dry conditions; the second effect results from the resistance to deformation of the tyre rubber when passing over the textured surface. If the horizontal traction force made available by a pavement to a foot or wheel is divided by the vertical force on that foot or wheel, the result is called the *coefficient of friction* of the pavement for that situation. Coefficients greater than one can occur in dry weather when strong molecular bonds associated with the first effect apply. In tests, greasy wood-block pavements were the worst of all pavements, with coefficients of friction of 0.2 when the wheel tyre was metal to 0.4 when the tyre was solid rubber.[771]

The frictional characteristics of a pavement surface can be improved by mechanically milling the surface by surface-dressing it with the aid of a bituminous spray (Chapter 14) or by overlaying it with a new rougher surface layer (Chapter 15). A drop in the frictional resistance of the surface due to traffic can be caused by abrasion, polishing or weathering of the surfaces of the upper layer of stones. External degrading factors are the effects of water (Chapter 2), oil or dirt. Greasy pavement surfaces were a problem for much of the initial stages of the motor-car age as car engines dripped oil and grease. These droppings accumulated at locations where vehicles were stationary, such as prior to the stop lines at intersections. Summer rains tended to spread the slippery grease and make braking dangerous at approaches to intersections.

The friction/skidding issue was recognised in 1906 in the UK, and early full-scale studies began at the National Physical Laboratory in 1927. The work was transferred to the newly formed Road Research Laboratory at Harmondsworth in the early 1930s and was continued at the Transport and Road Research Laboratory at Crowthorne. For directly assessing the properties of a pavement surface, initially researchers used a *sand patch* technique which recorded the amount of fine sand that it took to bring a defined area of a pavement surface to level. An early break-through occurred in 1931 when it was realised that skid resistance was very dependent on the weather.[772] Specific tests were developed to measure the texture of the surface and how much stones polish under traffic.[773] In 1952, the group modified an Izod pendulum device used for impact testing of metals to swing a standard piece of rubber across the pavement in question and measures the energy lost in the process. The results were empirically calibrated to align with skid resistance.

Methods using open-graded asphalt mixes to produce pavements with rougher surface textures were discussed earlier (Chapter 15). Paving techniques developed in the latter half of the 20th century to provide surfaces with good skid resistance were discussed in Chapter 14.

When a vehicle is travelling around a curve, a *centrifugal* force is produced which relates to the square of the vehicle's speed and applies outwards along the radius of the curve. This is resisted by frictional forces between the tyres and the pavement. If these forces are inadequate, the vehicle skids laterally off the pavement. Thus, the pavement must have good friction properties on curves. Safety is also enhanced by the pavement being cambered on curves so that contact forces normal to the pavement are also available to counteract the centrifugal force.

For vehicles travelling at automobile speeds, the propensity for a pavement to cause skidding or other losses of traction became more obvious. Road crash data consistently indicate that road crashes increase as pavement skid resistance drops at critical locations such as stop lines and curves. To overcome this problem, a number of devices for detecting low skid resistance, particularly of wet pavements, were developed. For turning traction, one UK device developed in the 1930s used a motor-cycle with sidecar mounted at an angle and measured the force between the two to provide a sideways force coefficient.[774] The motor-cycle was later superseded by in-board systems based on the same principle. For braking resistance, a common device used a trailer towed behind a car and measured the force in the tow bar between trailer and car when the trailer was braked and the wheel prevented from rotating. Although it could be used at highway speed, that use was restricted as the tyres soon overheated. The method was then employed in front-wheel drive cars and required a lead car to wet the pavement surface.

More recently the operating limits imposed on these devices led TRRL to the present-day commercial production of a sideway-force routine investigation machine (SCRIM) based in a truck with a freely rotating wheel moving at an angle to the vehicle direction. Water carried in the truck sprayed on the surface before contact by the locked testing wheel (Figure 19.5). Many factors, such as vehicle speed, affect skid resistance and so SCRIM must be used in a very standardised way if comparative date is required. Key external variations are ambient temperature and recent weather conditions.

In the mid-20th century, an increasing friction-related problem was occurring with traffic travelling in wet weather. If a film of water forms on the pavement, a vehicle can tend to slide along the surface of water if its speed is such that the water beneath the tyre does not have time to dissipate horizontally due to the forward speed of the vehicle. Instead, the water forms as a wedge between the tyre and pavement. The tyres lose any traction, and the vehicle can no longer brake or turn. This dangerous situation is called *hydroplaning*. It can be alleviated by two main measures. First, horizontal pavements are provided with sufficient crossfall to

Figure 19.5 SCRIM manufactured by WDM for use in New Zealand.[775]

permit water to quickly drain from the pavement. Second, a *porous friction course*, as described in Chapter 15, is used to allow surface water to drain vertically from the surface and into the voids in the course. Porous friction courses also help dissipate the noise generated by vehicle tyres and provide enhanced friction for braking and turning in dry conditions.

Good traction is one positive property that pavements provide for road users. A related feature is the provision of visible road markings to guide road users safely and productively along the road. Centreline marking to separate vehicles travelling in different directions on the one length of pavement began on narrow bridges where vehicles had limited ability to move laterally. In the 1820s, the English Bridge at Shrewsbury used white stones to separate oncoming traffic. Painted white lines required a continuous paintable pavement surface and, as discussed in earlier chapters, these did not eventuate until well into the 20th century. The first use was in Wayne County in Michigan which began painting white centrelines on bridges and bends in 1911.[776] Bituminous sealers in the longitudinal joints of concrete pavements were soon performing a similar function (recall that Wayne County was an early adopter of concrete pavements, Chapter 17). Line-marking machines were introduced in England in 1927. In 1937, Harry Heltzer, working for 3M in Minnesota, invented the reflective glass beads that make painted line-markings far more effective.[777]

The friction properties of a pavement related to factors with a very small longitudinal dimension. With the development and widespread adoption of high-speed pneumatic-tyred motor vehicles at the beginning of the 20th century, it became increasingly important to consider larger scale factors and produce pavements with a longitudinal profile that was sufficiently uniform to avoid making vehicle travel uncomfortable for people, damaging freight and being potentially dangerous.[778] Unevenness in the longitudinal profile of pavements was first observed in asphalt roads early in the 20th century.[779] The geometric consequences of this unevenness are called *corrugations* at a close spacing and *waves* at a larger spacing. A varying pavement longitudinal profile over a short distance (e.g. less than a metre) is commonly perceived by a road user as a pavement impediment that may need a steering manoeuvre or braking.

It soon became apparent that much of the deformation arose from a dynamic interaction between the pavement and the suspensions of passing vehicles. It was exacerbated by any vertical profile changes in the initial pavement and so improved construction practices could partially alleviate the problem. This particularly applies to corrugations, which are a series with an amplitude of over 10 mm and at relatively constant pitch. They are common on the approaches to curves on unbound pavements where vehicle braking often occurs. There is no simple way of preventing corrugations other than by providing a bound surface.

A long straight-edge could be used to measure the longitudinal road profile but it was a very slow, dangerous and tedious process, both to operate and to produce a useful output. The earliest known mechanical device was to Orograph which was produced for the American Army in 1853.[780] It used a long trough of mercury to provide a horizontal datum and then drew the profile on a paper roll. In about 1900, John Brown of Belfast and President of the Irish Roads Improvement Association produced his Viagraph. This was a 4 m long straight-edge with a wheel at one end and which also recorded the longitudinal profile of the pavement on paper from a paper roll.[781] The Viagraph was difficult both to operate and to then use its output.

More elaborate and effective profile measuring devices called profilometers were developed in America in the early 1920s, including a:

- tricycle-mounted rolling straight-edge in 1923 (Figure 19.6),[782]
- version of the Viagraph made by Hveem in 1929,[783]
- profilometer carried on 32 bicycle wheels developed by the Illinois State Highways Department and used for the Bates Road test in the 1920s (Chapter 20),
- walking profilometer produced by California Highways in the 1950s,[784]
- CHLOE[g] profilometer (Figure 19.7) produced by the Bureau of Public Roads for the AASHO Road Test (Chapter 20) in the late 1950s,[785]

[g] Named after its developers: Carey, Huckins, Leather and "other engineers" at the Bureau of Public Roads.

Figure 19.6 Tricycle-mounted rolling straight-edge, called a profilometer.[786]

Figure 19.7 Van pulling a Chloe profilometer.[788]

Figure 19.8 British Road Research Laboratory's 1960s hand-operated multi-wheel profilometer.[789] It closely resembles the Mailander Wave Measurer developed
by the Illinois Division of Highways.[790]

- British 1960s multi-wheel solution, as shown in Figure 19.8,
- GM profilometer produced in the 1960s and discussed below, and
- compact walking speed device developed by ARRB in the 1990s.

One major problem with these devices was how to maintain an absolute datum from which to measure the actual profile as a straight-edge spanning between two exceptional points would produce misleading data. To overcome this, the multi-wheel arrangements mechanically "averaged out" the high and low points encountered.[787]

Profile changes over less than about 3 m are also usually the result of pavement distress but may not be visible to the road user. Longer wavelength profile changes usually relate more to some subsidence and can be disconcerting to travellers and cargo, particularly when the vehicle suspension in some way resonates with the profile changes. The layman's *bumpy road* could embrace all these effects. Technically, as a group these profile changes are described as *road roughness*.

There was a growing realisation of the importance of roughness to travel in motorised vehicles and by the mid-1920s road roughness was being measured by wheeled devices with increasing levels of sophistication. The first of these was the Lockwood Roughness Integrator developed in the USA in 1923. It utilised the vertical motion of a moving vehicle, initially by measuring the vertical accelerations of the rear axle of the vehicle.[791] This was simplified in a 1925 version called the Relative Roughness Determinator which measured the relative movement between the vehicle body and the front axle.[792] New York State produced a similar device called a Via-Log[793] and another was developed in 1931 in Victoria (Figure 19.9). The Victorian vehicle was operated for about a decade until it began to produce inconsistent results and was replaced by a BPR Roughometer discussed below.

Figure 19.9 Roughness meter developed in Victoria, Australia, in 1930. The measuring device is the white box and vertical rod just to the rear of the radiator. The vehicle is a T-model Ford with helical front axle springs.[794]

To avoid the results being dependent on the characteristics of the vehicle, in about 1930 the American Bureau of Public Roads (BPR) standardised a device called the Dana Automatic *Roughometer*, which was mounted in a special standardised trailer towed behind a car being driven at about 30 km/h. It was further incrementally developed during the 1930s (Figure 19.10) and for some decades was the conventional device for measuring road roughness.[795] In 1946, the British introduced a similar device called a *bump integrator*. It was widely used with some 15 in operation in the 1970s.[796] There were two problems with these towed roughness meters: they could not operate at much over 30 km/h, and their data were not specific to a typical common wheelpath.

A return to vehicle-based roughness meters overcame both these problems. In 1967, Ivan Mays of the Texas Highways Department developed a vehicle-based device which measured relative movements between the vehicle body and the rear axle on a more user-friendly data-logger held on a passenger's lap. It was called the Mays Roadmeter (or Ride Meter) and was put into wide use by the American Portland Cement Association. A modified Roadmeter was produced by ARRB in Australia in about 1970 where it subsequently became an important tool for pavement management.[797]

Further incentives for development occurred with the appreciation of a strong link between user perceptions of pavement quality and measures of roughness. The devices had to be non-invasive, relatively quick and simple to operate and produce output that was reproducible and meaningful to

Figure 19.10 The BPR Roughometer for measuring road roughness.[799]

both experts and laypeople. The direct measurement of road roughness was found to have two direct benefits to pavement managers. First, it provided an acceptable measure of the quality of a newly constructed pavement and, second, subsequent measurements provided useful early indications of pavement deterioration and of the need for pavement maintenance. An interesting early construction lesson from roughness measurements was that the first length of a new pavement was always rougher than subsequent lengths, suggesting that it took some time for roadmakers to adequately control their working processes.[798]

By the mid-1980s, the availability of compact accelerometers and laser-based distance measurement allowed equipment to be developed which could be mounted in a conventional road vehicle travelling at highway speeds. This was mainly because there was no longer any need for physical contact between the device and the pavement surface. The output provided continuous profile data of sufficient accuracy to not only provide road roughness data but also measures of rutting and surface texture (Figure 19.11). The profile data were of sufficient accuracy to allow a computer model of a car – called a "*quarter car*" because it only used one wheel – to traverse the profile and give the same roughness estimate as would have been obtained by driving a vehicle-based roughometer over the same road. The quarter car was calibrated to match the characteristics of a late 1970s American passenger car travelling at 80 km/h.[800] An initial delay in the development of these methods was that more data were being collected than could be handled by the then-current computer systems.

Recent laser-based devices also allow the detection of cracking and deflection of the pavement surface. A modern pavement survey vehicle capable of

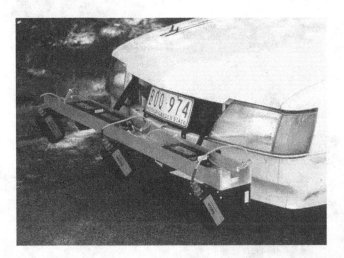

Figure 19.11 1985 ARRB profilometer. The measuring lasers were mounted on a transverse bar at the front of the vehicle.[801]

crack measurement at a road speed is shown in Figure 19.12 and its output is shown in Figure 19.13.

A significant finding of the AASHO Road Test to be described in Chapter 20 dealt with assessing the "serviceability" of pavements based on

Figure 19.12 The ARRB Road Survey vehicle, c2019.[803]

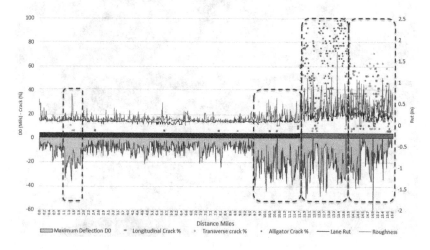

Figure 19.13 Pavement inventory output from a similar road survey vehicle, giving loaded deflection, three forms of cracking, rutting and roughness for 20 km of road.[804]

the premise that the road user should determine whether or not a pavement is satisfactory. It introduced a *Present Serviceability Index* (PSI) which was obtained by statistically correlating user opinions with road roughness (as measured then by the AASHO "CHLOE" slope profilometer (Figure 19.11) and the extent of cracking, patching and rutting.[802] During the test, a user panel drove over selected pavements and rated them using a scale from 0 to 5. A rating of zero denoted an impassable pavement whereas a rating of 5 indicated a perfect pavement. In addition, the raters were asked which objective features of the pavement influenced their rating and whether the road was acceptable for interstate traffic. The rating numbers assigned by panel members were averaged and designated the *Present Serviceability Rating* (PSR). Data presentation in a more positive light indicating smoothness rather than roughness is shown in Figure 19.14.

Later, a PSI was adopted by many paving engineers and by several American state highway departments for setting up maintenance programs, road life studies and work priority ratings. The Index was based on the same 5-point rating system as Rating but went beyond a simple assessment of ride quality. Its use was not a simple process and was mired by its statistical basis. Most road agencies now accept that quantitative roughness measurements are a reasonable surrogate for the views of road users without the need for an intervening PSR.

The various pavement roughness studies led to some wider understandings of the reactions of users to the pavement profile. For example, it was

Figure 19.14 Repurposed Australian version of CHLOE; developer John Scala is on the right.[806]

realised that vehicles do not respond to profiles with wavelengths greater than 30 m but are particularly sensitive to wavelengths of between 1 and 20 m; people seated in vehicles respond to vertical accelerations greater than 0.04 g and to vehicle vibrations of between 3 and 8 Hz.[805]

The availability of a pavement longitudinal profile led the World Bank to develop its now widely used *International Roughness Index* (IRI)[807] as illustrated in Table 19.4. A key advantage of the IRI is that it depends only on the longitudinal profile of the pavement surface and has no subjective component.

A series of international tests related the IRI to the various vehicle-based roughness measures discussed above. Thus, a practical advantage of the profile-based approach was that it permitted the various roughness measuring devices to be accurately calibrated without reference to the ride characteristics of the vehicle being used.

Such testing in conjunction with field testing of Brazilian pavements gave measures which were empirically related to future pavement performance, user perceptions and vehicle operating costs.[808] This allowed the Bank to promulgate more sophisticated approaches to optimising the nett benefit of road transport operations. In particular, the Bank promoted a life-cycle costing approach to pavements adding the life-cycle benefits from the use of the road to its costs of construction, maintenance and rehabilitation. The approach relies on a statistical model using field data to predict roughness from key variables such as pavement age and strength, traffic usage, environmental conditions, maintenance practices and the local environment. It was incorporated into the Bank's Highway Design and Maintenance Model (HDM).[809]

The model was further developed to make it transferable to other regions[811] and the subsequent versions are now managed by the international road organisation, PIARC, to provide a common decision-making tool for checking the engineering and economic viability of the investments in road projects. As a consequence of this work, *pavement performance* which had been the sole domain of the pavement technologists over time became subsumed into *asset management*, particularly as the road agencies

Table 19.4 Roughness examples using the IRI measure[810]

Pavement type	Description	Typical speed (km/h)	IRI
Airport runway, super highway	Very smooth	100++	0–2
New pavement	Smooth	100+	2–3
Old pavement	Surface imperfections	100	3–6
Maintained unpaved road	Frequent minor depressions	100	5–10
Damaged pavement	Frequent shallow depressions	80	5–10
Rough unpaved road	Erosion gulleys, deep depressions	50	10–20

became road managers rather than road builders and broadened their interests beyond *project management* to *network management*. In the process road roughness came to be seen as a *functional response* and pavement strength as a *structural response*.

Increasing experience has continued to show that in most cases road roughness is the best indicative measure of pavement performance. Nevertheless, there are circumstances where rutting, cracking, climate or surface polishing may be better performance indicators than roughness. However, for pavements that have been appropriately designed, constructed and maintained for the traffic that has subsequently used them, pavement age is closely linked to all the above indicators and so performance models simply linking roughness to pavement age have wide application in pavement management methods.

Of course, the inherent variability in pavement materials, construction processes, traffic and environmental conditions and inter-relationships between these properties and with pavement age means that performance predictions must have a strong statistical basis and be couched in clear statistical covenants. In 1996, an expert review concluded[812] that *road roughness was the most appropriate means of assessing long term pavement performance because it was an objective measure, inexpensive to collect, related directly to user costs and was the most relevant measure of the long term functional behaviour of pavements.* The advice was wise in part but ignored the need to also visually monitor the integrity of the pavement, particularly with respect to its water-tightness.

Pavements can be supplied and maintained via specifications which provide either: (1) a detailed *recipe* that the supplier must follow, with the customer carrying the risks of the recipe producing an inadequate product or (2) a *performance* specification where the supplier carries the risk that the pavement it produces will behave as the customer requires. The tortuous nature of pavement development during the 19th and 20th centuries meant that the distinction between the two was often blurred.

The work of Telford and McAdam early in the 19th century (Chapters 8 and 9) saw the British Government and turnpike owners inherently accept that the pavements of Telford and McAdam would perform as expected throughout their economic life. The confidence spread to America where recommended pavement construction contracts in the pre-asphalt 1840s were entirely recipe-based with no mention of subsequent performance.[813] However, at that same time the advent of asphalt paving in Paris (Chapter 13) saw many of the new pavements fail very early in their service life. Pavement owners in Paris and London began ask paving contactors for performance guarantees. This move subsequently bankrupted most contractors. Early in the 20th century, asphalt technology had improved to the extent (Chapter 15) that American contractors were offering asphalts with precisely patented recipes and somewhat less precise performance guarantees. Nevertheless, in 1968, an American Federal Highway Administration report notes that *an ideal quality assurance system (would have) the owner describing what is wanted by design drawings and specifications* with no mention of subsequent pavement performance.[814] Contractually based performance guarantees developed gradually later in the 20th century, aided by the progressive development of the objectives measures of pavement performance discussed earlier in this chapter.

The way in which a pavement performs is very dependent on its design, the materials used, how it is constructed and maintained and how it is used. The discussion in the earlier parts of this chapter focussed on the individual components of this quintet. Building on earlier methods of quality control, a major set of changes that occurred in the 1980s related to a better recognition of the statistical variation of all these factors, particularly in the context the new processes of total quality management (TQM). A core requirement of TQM was for the pavement owner to clearly define how it would determine whether a pavement was fit for its intended purpose. This was managed by both process control during production and acceptance control after production. Importantly, the obligation for testing was on the producer, although the consumer could audit the processes and results. In retrospect, an important part of TQM was for any factor negatively impacting on the pavement being fit for its defined purpose (called a non-conformance in TQM jargon) to immediately be formally reported and acted upon. TQM delivered many benefits for pavement producers and pavement owners, particularly if it also involved an equitable distribution of risks between the two parties.

Underlying pavement TQM was a realisation and acceptance that all pavement construction and maintenance processes would be of variable quality, despite the best human intentions. This led to contracts between consumer and supplier specifically defining an acceptable proportion defective and a sample size for each product. For example, a key property in pavements is the pavement thickness which, not surprisingly, is shown by post-construction measurements to be in statistical terms *normally distributed*. The following data show the real-life variation pavement thickness on some "good" jobs.[815] Note that a *standard deviation (sd)* means that 15% of the results would have been further from the intended thickness than the sd.

Pavement placed by laser-controlled equipment	sd = 10 mm
Pavement placed by best quality (pre-laser) equipment	sd = 12 mm
Pavement placed by conventional grader, with some survey equipment	sd = 15 mm
Pavement placed by grader, using eye and closely spaced pegs	sd = 25 mm
Pavement placed by grader, using eye and widely spaced pegs	sd = 30 mm
Pavement placed using non-professional staff	sd = 45 mm

Of course, there is also the issue of where and how often thickness is measured along a pavement. The fact that most of the properties of a constructed pavement could vary in a major way from the specifiers' assumptions first became clear during the AASHO Road Test (Chapter 20) in the early 1960s. Clearly, life in the real world of pavements is not as precise as some of the earlier theoretical approaches might have implied.

The first step in the 1960s was to so-called end-result specifications which measured completed pavement in statistical rather than absolute terms. At this stage, users could still only speculate on the subsequent performance of the pavement. The next steps required ways of technically predicting pavement performance, and of then commercially relating payments to future performance. The ultimate aim is to optimise the total nett benefits of the pavement over its life. These issues are beyond the scope of this book but have been addressed elsewhere, particularly by Chamberlin for the American Transportation Research Board in 1995[816] and via the experience of toll road operators who must balance over the life of a project its construction costs, maintenance and replacement costs, and users' perceived benefits.[817]

In the 1970s, many of the above processes were linked by formal *pavement management systems* (PMS). These bring together pavement planning, design, programming, construction, operation and maintenance into a single predictive model with the aim of optimising the costs of pavement construction, maintenance, rehabilitation and reconstruction within a defined level of service. If a pavement's condition was less than required, an optimum solution to its restoration was sought. Typically, this would result in prioritising forward programs for work on a pavement progressively involving routine maintenance, preventative maintenance, rehabilitation and finally reconstruction. For example, this could indicate crack sealing every 5 years, full surface sealing every 8 years, an overlay every 15 years and reconstruction after 30 years. To be effective, pavement management systems require continuous feedback of all the costs and benefits associated with providing and operating a pavement.

Pioneers in the formal development of PMS were Ralph Haas and Ron Hudson. In their important book[818] they observe:

> Pavement management was born in the mid-1960s largely in response to numerous unanticipated pavement failures on the US Interstate and Canadian Highway Systems. These roads had been designed and constructed using the best known pavement design technology at that time, including the results of the AASHO Road Test. After an intensive national review of problems observed, the impossibility of making accurate single-point predictions of pavement performance due to national statistical variability of the major inputs became clear. Design methods at that time required as inputs estimated traffic, projected as-constructed materials properties, and estimated environmental

conditions for a 20–30 year life of the pavement. These methods did not take into account the effects on performance of pavement maintenance, nor did they consider the life-cycle cost past the initial design period to include one or more overlays and rehabilitation activities, which everyone knew were common practice on heavy duty pavements.

They then noted that it was essential to measure pavement behaviour since known theoretical pavement equations did not predict pavement performance directly. They pointed out that it was possible to determine the required maintenance, overlay, and rehabilitation needs of a pavement including expected time of such interventions using PMS to optimise total life-cycle costs.

Given all the factors discussed earlier in this chapter, a general overview of pavement life is shown in Figure 19.15. Overall, building extra life into a pavement at construction is usually cheaper than frequent maintenance and rehabilitation interventions, particularly if road-user costs are part of the evaluation.

Between the two extremes indicated in Figure 19.15, the various pavement performance measures discussed above help pavement managers deliver the service that road users desire, as outlined at the beginning of the chapter. As the Figure shows, pavements deteriorate with time and measures to help managers plan the remediation of the pavement range from the rapid closing of surface cracks and repairs of potholes, to major long-term interventions to replace or enhance the pavement structure. Priority is often given to crack sealing as a fore-runner to other problems, particularly as water entry into a basecourse via cracks will often seriously degrade the strength and stiffness of the basecourse.

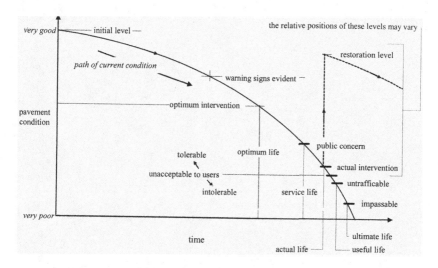

Figure 19.15 Pavement manager's view of the life of a pavement.[819]

An important aspect of pavement maintenance can be demonstrated by analogy to maintaining the paint on the outside of a building. The stich-in-time philosophy applies and it is usually cheaper in the long term to repaint the walls before the paint begins to peel away. The same applies to pavements, particularly where their functioning depends on minimising water entry. Unfortunately, many of the roughness-based methods described above lead to interventions at the paint-peeling stage. While such interventions are needed, it is important for road managers to also employ frequent lower cost maintenance interventions, even on pavements with a low roughness reading.[820]

Chapter 20

Full-scale testing of pavements to validate designs using local materials

From the earliest times, the performance of pavements would have been monitored in some way or other and the lessons learnt applied in the building and maintaining of new pavements. As new technologies became available, some material testing – as has been discussed in Chapters 4 and 19 – might have preceded field tests intended to evaluate the overall performance of the as-built pavements. However, it was not until the advent of the motor car that testing of the materials used and of the product produced became essential. Increased patronage brought increased scrutiny of pavement performance. And yet, as we have seen in the preceding chapters, the real world of pavement design, construction and management was far from being neat and precise. Approximations, empiricisms and misconceptions abounded, and cost-cutting was applauded.

An obvious first step away from laboratory tests and preconceptions was to observe how real pavements behaved under real, or almost real, traffic.[821] Such full-scale tests were, more or less, conducted under controlled conditions and known traffic loads and, although expensive, they had the potential to provide useful new information. A prime practical purpose of such tests was to assess the performance of a pavement before it was put into service in the real world. Some were conducted on pavements which were lengths of functioning roads and some on pavements built in the field but independently of the real road system. A third step was to test specially constructed pavements in laboratory conditions.

At the beginning of the 20th century, the relative cost-effectiveness of laboratory tests, tests on real pavements and special pavement test facilities was examined by Crompton in Britain. His analysis generally favoured tests on real pavements.[822] Thus, the building of experimental pavements to carry normal traffic on selected public roads was an approach mainly used in the UK.[823] From 1930 to 1966, some 400 test sections were installed on public roads at the sites listed in Table 20.1[824] and subjected to everyday traffic. Some of the sites were under observation for up to 60 years. Their primary purpose was to develop designs for both asphalt and concrete roads and then to produce the best "recipe"

Table 20.1 British pavement test roads

Year	Location	Type	References
1911	Sidcup, Kent (A20)	Tar macadam	Boulnois (1919, p140)
c1912	Kingston Vale, London (A3)	Tar macadam	Boulnois (1919, p141)
c1913	Fulham, London, "streets"	Tar macadam	Boulnois (1919, p142)
1930	Great Chertsey Road (A316)	Concrete	Croneys (1997, p302)
1931	"trunk road"	Tar macadam	Hadfield (1934, p113)
1932	Kingston Bypass (A3)	Asphalt	Hadfield (1934, p115, 142); Hosking (1992, p161)
1933	Hampton Court Road (A309)	Concrete	Croneys (1997, p302)
1937	Colnbrook Bypass (A4)	Bitumen	Road Research Laboratory (1962, p180)
1940-44	Colnbrook Bypass (A4)	Tar	Road Research Laboratory (1962, p181)
1945	Colnbrook Bypass (A4)	Bitumen	Road Research Laboratory (1962, p180)
1945–1949	New Kent Rd, London	"Dense tar"	Road Research Laboratory (1962, p281)
1946	Oxton, Notts (A6097)	Concrete	Croneys (1997, p302)
1947–1954	Upper Richmond Rd, Putney	Asphalt	Road Research Laboratory (1962, p282)
1948	Longford to Stanwell (B379)	Concrete	Croneys (1997, p302)
1948	Huntingdonshire (A1)	Asphalt	Road Research Laboratory (1962, p283)
1949	Boroughbridge, Yorksh (A1)	Asphalt	Road Research Laboratory (1962, p225)
1951–1953	Haddenham, Buckingham	Surfacing	Hosking (1992, p62)
1952–1955	Streatham High Rd (A22)	Surfacing	Hosking (1992, p149)
1955	Llangyfelach, Glams (A48)	Concrete	Croneys (1997, p303)
1956	East Retford, Notts (A638)	Asphalt	RRL
1957	Alconbury Hill, Cambsh (A1)	Asphalt	Croneys (1997, p276)
1957	Alconbury Hill (A1)	Concrete	Croneys (1997, p303)
1961	Winthorpe, Notts (A46)	Concrete	Croneys (1997, p303)
1962	Grantham Bypass, Lincs (A1)	Concrete	Croneys (1997, p303)
1963	Nately Scures, Hamps (A30)	Asphalt	Croneys (1997, p276)
1963	Wheatley Bypass, Oxfs (A40)	Asphalt	Croneys (1997, p277)
1964	Alconbury Bypass (A1)	Asphalt	Croneys (1997, p277)
1965	Conington, Cambsh (A1)	Asphalt	Croneys (1997, p277)
1966	Tuxford Bypass, Notts	Concrete	Lock (1968), RRL Rep LR 218

for use in future road-making.[825] From 1965, the program became the basis for British pavement design practice for new roads via Road Note 29[826]. As will be evident from the rest of this chapter, this slow real-time approach was the opposite to the accelerated testing adopted in most other countries. Both approaches produced useful pavement designs for their host countries, but this chapter will later suggest that neither outcome was optimal.

It is not a coincidence that three of the test roads listed in Table 20.1 were on the Colnbook Bypass, about 25 km west of central London. The Bypass opened in 1929, and in 1930, the British Ministry of Transport opened a road-testing laboratory at Colnbrook. In 1933, it was taken over by the Department of Scientific and Industrial Research and formed the nucleus of the Road Research Laboratory, which later transferred to Crowthorne, 30 km to the south-west.

While the observation of the performance of real roads had many merits, it also suffered from three main disadvantages. First, it was inherently retrospective, and the results were often subsumed by the rapid growth in traffic. As was subsequently observed:

> there were considerable increases in the damaging power of commercial vehicles using the major road network and the most heavily trafficked roads are now designed for cumulative traffic of up to ten times that envisaged when the first motor-ways were opened. The consequential increase in need to repair older parts of the network led to a major increase in reconstruction work, which presented distinctive design problems, particularly as there was also interest in designing for longer lives in order to reduce future maintenance and the associated traffic delays.[827]

Second, the real roads tested reflected then-current conditions and did not explore future conditions and pavement-making techniques. Third, as noted in earlier chapters, there was a real need to better understand how new pavements would perform over time, particularly because the understanding of pavement behaviour was still far from satisfactory (e.g. see Chapter 16). A British review in 1984 commented:[828]

> Laboratory tests that seek to simulate conditions in the road pavement are gross simplifications and although they yield results that are generally suitable for characterising the dynamic stress and strain behaviour, prediction of long-term performance from them is much less certain. Simplifying assumptions have also to be made in developing these models; for example, the materials of the pavement are treated as homogeneous, isotropic layers of idealised material. Consequently, the many complex models that have been developed stretch structural theory in seeking to characterise the inhomogeneous particulate materials used in road construction.

The alternative to testing real roads was to test full-scale pavements. The first full-scale pavement test was conducted in 1816 by a committee of the Dublin Society under the leadership of Richard Edgeworth. The test showed that a broken stone pavement required much lower haulage forces than the pre-macadam alternatives. Edgeworth died in the following year, and the pavement test is not mentioned by his biographer.[829] Tests were conducted in France in the 1830s by Claude Navier and Gustave Coriolis of the Corps des Ponts et Chaussées. The mid-19th-century tests of Morin in France and Telford and associates in England to determine the haulage demands of various pavement types have previously been mentioned (Chapter 13). Haulage forces were a very important pavement attribute in the era of animal-hauled vehicles. Another early concern of pavement engineers was the wear of stone or brick pavements caused by the horseshoes and steel-rimmed wheels (Chapter 10) of horse-drawn traffic.

The next step in the testing of real pavements was to move from relatively uncontrolled traffic loading to artificial but well-defined loadings. It was not a simple step as each moving wheel load had to be at least 2 ton, requiring quite large machinery to move a wheel applying such a known large load along a pavement for a known number of passes. After each pass, the change in the pavement would then be measured using the methods, as described in Chapter 19. The first planned artificial loading device was a trafficking machine designed by Irish engineer RJ Kirwan in 1905 (Figure 20.1). Apparently, it was never built.

Kirwan's intended device was followed in 1909 by a circular pavement test track built in Detroit by JC McCabe, a boiler inspector in the city's Department of Public Works Detroit city.[831] It was called the *paving determinator* and was specifically developed to test *the quality of paving materials under heavy abrasive influences*. It employed a 3 m radius circular test track with eight 500 mm wide test pavements placed along its circumference (Figure 20.2). One of its circulating wheels carried five simulated hooves and horseshoes, and the other was a steel-rimmed cart-wheel to simulate the then-current vehicular traffic.[832] The wheels weighed 640 kg, and the "horse shoes" had a contact pressure of 1,000 kPa (about 30% higher than a modern truck tyre). The wheels were driven at speeds from 5 to 20 km/h, although testing was usually at about 11 km/h. An eccentric

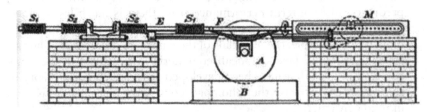

Figure 20.1 Kirwan's 1905 linear trafficking machine. [830]

Figure 20.2 The paving determinator.[834] The hoof loading wheel is at the left end of the revolving diametric arm, and the iron-rimmed wheel is at the right end.

drive mechanism moved the wheels radially across the strip.[833] The device quickly showed the difference in the performance of various brick, stone and concrete surfaces. For instance, the macadam surface failed under the horseshoes but otherwise became well compacted. The more general purpose of the tests was that, for reasons discussed in Chapter 17, Michigan was building America's first concrete-paved rural highway on a mile length of Woodward Avenue just outside the Detroit city limits. The chosen design was "validated" by the Determinator's full-scale tests.

As a consequence of the demands caused by the US entry into World War I, their railroads were unable to cope with the increased demands and heavy motor trucks came into wider use but – as discussed in Chapter 19 – their increasing loads and frequency during the war caused *many heretofore suitable roads* to be *completely destroyed*. In response to this, in 1918, the then Bureau of Public Roads (BPR, now Federal Highway Administration, FHWA) launched a set of road tests at the Department of Agriculture's Experimental Farm in Arlington, Virginia to study the four factors that then seemed important in the pavement design: (1) impact forces, (2) pressure from the passing loads, (3) horizontal shear and tractive forces and (4) subgrade and soil conditions.

A 55 m diameter circular test track was constructed with a 4 m wide test pavement. The pavements were loaded by trucks travelling at about 20 km/h. The wheel loads were about 2.5 ton, but this was later increased to 3.5 ton[835] (Figure 20.3).

Figure 20.3 BPR test track in about 1925.[837]

The initial tests modified the design of concrete pavements and, in particular, were used to establish the required layer thickness.[836] Later tests were on asphalt pavements and looked at asphalt's temperature sensitivity and at the role of air voids. Overall, the tests were relatively inconclusive, although they did demonstrate some very major shoving and rutting of the asphalt pavements being tested.

Shortly after this circular track was developed, within the track the BPR began to simulate the effect of traffic loads by means of impact devices, the first simply dropping a weight onto the pavement, followed by attempts to use a wheel supported by a steel frame (Figure 20.4). Later extensions of this approach are shown in Figure 20.5.

Companion testing to the impact loader was conducted using chassis-based wheel loads, as shown in Figure 20.5. Note that the loads were still being applied via solid rubber tyres.

In 1911, experience with the British trial road sections (Table 20.1) led to the suggestion that similar information could be obtained more rapidly *by means of a machine which should apply to samples of various surfacing material the same sort of stresses as took place on the road... the result was... the 'Road Machine'.*[839] The 'Road Machine' was developed at the British National Physical Laboratory (NPL) at Teddington in southwest London in 1911–1912.[840] The circular track with a 10 m diameter was loaded by eight steel-rimmed wheels, each independently driven, on radial arms and operating at 5–13 km/h (Figure 20.6). The wheelpaths were varied by a cam mechanism to impart a transverse motion to the wheels, the load on each wheel could be adjusted and a hot air duct system,

Figure 20.4 The BPR impact loader.[838]

Figure 20.5 The BPR machine for applying wheel loading to a pavement via a chassis-like frame (1924).[842]

Figure 20.6 The 1912 National Physical Laboratory road-testing machine.[843]

discharging in front of the wheels, could alter the road surface temperature. Instrumentation was also developed to measure transverse and longitudinal profiles. The intention was to simulate 1 year's wear on a typical heavily trafficked road in 24 hours.[841]

The initial four tests were on macadam, tarmacadam and two penetration macadams using two different pitches as binders. The first three failed prematurely. The next five tests were pavements with 40 mm of various asphalts placed on 150 mm of concrete. Two performed well and all emphasised the need for compaction. In 1916, the machine was temporarily retired until 1920. An overall conclusion from the early work is still valid and stresses the importance of construction practice:[844]

> An important finding from the road machine tests was the importance of achieving a low voids content in asphalt surfacing, this was corroborated by experience with five full scale experimental roads. It was also pointed out that inadequate bitumen content, the type of sand and the initial compacted density were critical factors for good performance. (see Chapter 15 and Table 20.1)

In 1930, the NPL machine was transferred to the Road Research Laboratory at Colnbrook/Harmondsworth, and replaced in 1933 by a 34 m diameter track with loading was applied by a full-sized 'captive' lorry with dual pneumatic-tyred rubber wheels (Figure 20.7).[845]

Experiments on test roads increased soon after the transition from horse-drawn to motor vehicle traffic during World War I. A 10 km controlled-access highway called Avus (*Automobil-Verkehrs und Übungsstraße* – automobile traffic and practice road) was built in Berlin, starting in 1913. It was originally built as a race-track but was also used for testing vehicles and road surfaces.

Figure 20.7 The 1934 RRL "Road Machine No 3". The iron-rimmed tyres in Figure 20.6 have been replaced by dual pneumatic tyres.[846]

An elliptical test road sponsored by Columbia Steel Company was built in Pittsburgh, California in 1921. The test loops were built off the roadway between Pittsburgh and Antioch, California. It was intended to determine the benefits of using reinforcing steel in concrete pavements (Chapter 17). Thirteen instrumented test sections from 130 to 200 mm thick were used and examined different cross sections (including thickened edges), amounts of reinforcing and the location of longitudinal joints. Loads were applied by surplus World War I army trucks with solid rubber tires. The studies showed the need for control of wheel loads and led to the use of an 8 ton standard maximum axle load (2 ton per tyre, Figure 16.5) and the phasing out of solid rubber truck tyres by 1926.

In 1920–1923, Illinois operated the Bates Test Road to find a cost-effective way of meeting local demands for better roads, as discussed in Chapter 17. The 3 km test pavement ran from Bates (SW of Springfield) east to the Farmingdale Road on what is now the old US54. There were 71 test sections made with different materials and to different designs. They were driven over by surplus World War I army "Liberty" trucks with dual 125 mm solid rubber tires on their rear axles, with loads ranging from 1 to 6 ton (Figure 20.8). The tests showed that of the 22 brick, 17 asphalt, and 32 concrete sections in the experiment, just one brick, three asphalt, and ten concrete sections satisfactorily withstood the imposed traffic loadings. The excellent performance of the concrete pavements led engineers to favour the use of longitudinal centre joints (Chapter 17). The results were also used to develop an equation relating required pavement thickness to traffic loading based on the theory of cantilever beams. In 1926, highway officials

Figure 20.8 Bates Road Test (1922–1923).[848]

estimated via the Associated Press that the pavement tests saved the State $9 million in construction costs.[847]

Many other countries also developed test tracks, with a track operating in Bandung in Indonesia in the late 1920s (Figure 20.9). It was developed in co-operation with Dutch engineers and used two tethered vehicles, a truck and a two-wheeled goods cart (a *grobak*), which ran around a 175 m long oval test pavement. The pavements were 5.5 m wide. This idea has recently been developed further in Spain at the Centro De Estudios De Carreteras Test Facility (CEDEX).[849] In 1928, an outdoor 20 m diameter circular test track began operating at Karlsruhe University in Germany.[850]

Pneumatic-tyred vehicles were now being widely used. The Bois de Vincennes – Maison Blanche test track was built in south-east Paris in 1931 by Victor Legrand at the initiative of the Ministry Public Works, the Ministry of War and the City of Paris. The decision to build it was motivated by the fact that *the study of the problems relating to the constitution and resistance of the pavements [...] had previously been conducted in France, either through research carried out in closed laboratories or on pavements open to general traffic [ ] where the experimenter was not in control of the critical factors....*[852] The track was substantially a quadrilateral, 930 m long and 6.5 m between curbs. It had a Trésaguet-type foundation of large vertical stones (Figure 8.3) on which the test pavements were placed. The first tests were aimed at comparing the wear of macadam pavements covered with different types of superficial coatings, under the effect of 7 ton trucks, one with solid rubber tires and the other with pneumatic tires. The latter, of course, came out the winner! They also used two 10 ton test vehicles, travelling at either 30 km/h or at 60 km/h, on different types of pavement, including tarmac, asphaltic concrete, concrete, blocks, etc. The tests focused on the wear of the surfacings rather than on the design and dimensioning of the pavement.

A laboratory-scale test track built by the Bureau of Public Roads at Arlington, Virginia in the 1930s and supplemented the larger track discussed earlier (Figure 20.6). It tested a circular pavement with a mean diameter of 3.6 m and built in a 450 mm wide and 310 mm deep concrete trough.

Figure 20.9 Test track at Bandung, Indonesia.[851]

Two wheels with a weight of only 350 kg on low pressure tyres were used to compact and test the pavement (Figure 20.10). The speeds ranged from 6 to 14 km/h. The wheels could be distributed across the width of the track to simulate traffic.[853] The tests reported in 1936 were on sheet asphalt or oiled broken stone surfacing (Chapter 14). They demonstrated the need to control the proportion of very small particles in a pavement mix.[854] Subsequently, similar installations were built in many other countries. A review of their output concluded that they had *been cost-effective in generating significant developments in road pavement technology.*[855]

From 1944 to 1954, the BPR tested asphalt pavements using newly developed loading methods and equipment to measure vertical displacements of the pavement. The test site was a 700 m oval built at Huntley Meadows Park near Hybla Valley in Virginia. The loaded, pneumatic-tyred wheels of the testing machine bear on the pavement and were guided by rails at the side of the experimental sections.[856]

Related work in conjunction with the Highway Research Board (which became the Transportation Research Board in 1974), several States, truck manufacturers, and other highway-related industries were part of Road Test One – MD (or Maryland Road Test) in 1949–1950. An 9-year-old 2 km length of two-lane pavement at La Plata in Maryland was carefully inventoried, instrumented, and then traversed by 1,000 trucks per day. The results showed the value of good load transfer between pavement layers, the effects of speed and axle loads, and the problems caused by subgrade pumping (Chapter 1). It produced the first links between wheel loads measured when a truck is stationary and the equivalent (dynamic) wheel loads produced when the truck is in motion.[857]

The first full-scale laboratory test track in continental European appears to have been a circular track built in Iassy, Romania in 1952 followed by one in Germany in 1955.[859] The first Australian attempt was conducted by its national scientific organisation in 1960–1961, but no details of this

Figure 20.10 1930s Arlington test track.[858]

Figure 20.11 Australian pavement loading machine (c1960).[864]

facility, other than the photo in Figure 20.11, have been located to date. The first Australian test road was built by the (then) Main Roads Department in Barcaldine, Queensland in 1960.[860] It demonstrated the effectiveness of bituminous stabilisation (Chapter 18) of the local clayey "black" soils.

Between 1951 and 1954, the Western Association of State Highway Officials, a division of 11 western American States within AASHO, joined with BPR and the Highway Research Board to conduct a WASHO road test to establish load limits for asphalt pavements. It used two four-lane road loops built in Malad City, Idaho, over a subgrade of silty clay. The test gave information about the stresses in flexible pavements under varying truck loadings, but the results were confounded by local freezing and thawing conditions (Chapter 4).[861] It was found that rutting was due to the horizontal lateral and longitudinal movement of the overloaded material.

Indoor linear test tracks have been built by several organisations at various scales. In 1957, a one-third scale linear track was operating at 30 km/h over about 4 m at Washington State University at Pullman.[862] By 1963, there was also a 25 m diameter circular track. Sydney University operated a quarter-scale scale linear track in 1967–1970 (Figure 20.12).[863] It used a slowly moving wheel with wheel pressures of 600 kPa to measure the dynamic load of the wheel, the permanent horizontal and vertical movements of the pavement surfaces, and the transient deflections of the pavement surfaces under load. The tests indicated that vertical wheel loading could also cause horizontal deformation of a pavement. It also showed the importance of compaction of a basecourse and the value of stabilisation (Chapter 18).

Figure 20.12 Sydney University test track in 1970.[865]

Later studies were dominated by the very large American Association of State Highway Officials' (AASHO) national road experiment in 1958–1960. The project was conducted in Ottawa, Illinois on six two-lane test road loops on the future alignment of IS 80 (Figure 20.13). The pavements were loaded by a fleet of pneumatic-tyred trucks with different wheel and load configurations. The body managing the AASHO Road Test also included the key industry associations. Performance was measured by recording deflection, roughness and visible distress. The information obtained was crucial in advancing knowledge in a range of areas, particularly pavement structural design and performance, load equivalencies and climate effects. It resulted in the performance equations and nomographs used to develop a new pavement design guide, first issued in 1961 as the *AASHO Interim Guide for the Design of Rigid and Flexible Pavements*. A key finding from the tests was that a single layer of asphalt placed directly on the subgrade would perform adequately, provided it was of sufficient thickness. Further findings are discussed in Chapter 15. The test also introduced and propounded the Present Serviceability Index concept, as discussed in Chapter 19. This approach has been criticised as diverting attention away from causes and towards user-perceived consequences.[866]

In 1970, a one-third pavement testing device was installed at Washington State University.[867] It was similar to the UK Road Machine (Figure 20.4) and consisted of a 14 ton steel frame and water tank rotating on a 26 m diameter circle at speeds up to 70 km/h. Three dual wheels could each be loaded to 5 ton. Some of the tests related asphalt pavement on silty subgrades.[868,869]

Figure 20.13 AASHO Road Test in Illinois in 1958–1960. Surface view at the top and overview at the bottom.[870]

Loading equipment to fully simulate traffic loads on existing, or purpose-built, pavements was first developed in South Africa by its national scientific body (CSIR).

In 1971, it built a heavy vehicle simulator (HVS) intended for use on in-service pavements by applying single or dual wheels loaded up to 10 ton, driven to and fro by a hydraulic motor, over an 8 m length of pavement. The first stationary HVS was designed to simulate the damage done to airport runways, resulting from the impact of aircraft landing. It was built from Bailey bridge components and subsequently was known as the HVS Mk I. The force applied to the pavement came from water tanks placed above an aircraft wheel supported by the Bailey bridge structure. It produced useful results but its lack of mobility led to a mobile version that could test real road pavements (Figure 20.14).[871]

The HVS has been constantly developed over the last 35 years. There are currently about 18 units operating worldwide, and a large knowledge base has been developed with the data shared by HVS users through the HVS International Alliance. It was reported in 2019 that the benefit/cost ratio for projects was between 4 and 10, citing studies related to:[873]

- the effect of water damage on pavements,
- rutting control,
- benefits of deep and shallow pavement structures
- use of cement stabilisation,
- an improved mechanistic design method,
- new asphalt overlays to improve road performance,
- environmentally friendly construction,
- improved aggregate skeleton in hot-mix asphalt,
- new crushed-rock base material developed and approved nationally, and
- improved specifications on all aspects of pavement design.

Denmark has a pavement testing machine located in the Danish Road Institute in Roskilde (Figure 20.15). Holland operated one of the first full-scale in-door tracks in 1991 at its LINTRACK in Delft (Figure 20.16).[874]

Figure 20.14 South African heavy vehicle simulator.[872]

Figure 20.15 Danish road-testing machine (Danish Road Institute). [875]

Figure 20.16 LINTRACK, Delft University of Technology.[876]

An indoor circular test track with a 12 m diameter was built in Regina, Canada, in 1977 and a 32 m track in Zurich in Switzerland in 1978.[877] In the early 1980s, a circular testing machine, the "Manège de Fatigue" (Figure 20.17), was built in Nantes, France for the Laboratoire des Ponts et Chaussées and began operating in 1982.[878] The track has an effective diameter of about 35 m, and the pavements are about 6 m wide. The loads that can be applied are between 6 and 13 ton in single or tandem wheel configurations, applied at up to 100 km/h. The track was modernised in 2013. The facility has allowed France to refine its catalogue of pavement designs and improve its models for predicting pavement damage, with a particular focus on fatigue damage and rutting resulting from truck traffic.[879] It remains very active, and about six pavements have been tested each year.

In the mid-1980s, the growth of truck traffic on Australia's very large but sparsely supported road network was accelerating, and the need for some national effort to improve understanding of performance of relatively low-cost pavements became urgent. The Australian Road Research Board (ARRB) therefore decided to construct and operate a testing machine which was called the accelerated loading facility (ALF). The detailed design and construction were undertaken by the (then) NSW Department of Main Roads. The rolling wheels are driven by wheel mounted electric motors with the assembly running on steel rail. They apply loads over a 12 m length of pavement when moving in the traffic direction and then lift the load free of the pavement when returning in the opposite direction (Figures 20.18 and 20.19). Wheel loads of 4–8 ton are applied via single, twin or triaxle suspension arrangements.

ALF was 'launched' in Sydney at the PIARC international roads conference in 1983. It has since been in continuous use for 35 years and proved of great value. For instance, it demonstrated the importance of compacting unbound (macadam) pavements and the value of using polymer-modified bitumens (Chapter 14). An independent economic study[881] showed ALF to have yielded an overall benefit-cost ratio between 4 and 5, with individual

Figure 20.17 Laboratoire des Ponts et Chaussées' (now IFSTTAR) test track at Nantes, France.[880]

Figure 20.18 & 20.19 ARBB's ALF inside and on the road (Hume Freeway, Benalla, Victoria).[882]

projects ranging between 1.4 and 12. The assessment did not include road user benefits. It also showed that ALF had been a catalyst for pavement research and innovation and had improved co-operation between researchers and practitioners. Two ALFs were sold to the American Federal Highways Research Laboratory, one to the Chinese government and one to the State of Louisiana.

Figure 20.20 BASt pavement loading machine in 1995.[884]

More recently, two further test roads have been built and operated in the USA. MnROAD was a pavement test track owned and operated by the Minnesota Department of Transportation. It examined various potential pavement materials. The track was constructed between 1991 and 1993 with over 50 unique test sections on a 10 km main road segment and a 4 km low traffic segment. In 2000, the 4 km National Center for Asphalt Technology Pavement Test Track was built in Alabama. It is an oval-shaped track used for testing experimental asphalt pavements.

An indoor approach developed by the Germany federal road research institute (Bundesanstalt für Strassenwesen (BASt)) enabled the effects of impact loading to be examined (Figure 20.20).[883] Interestingly, in 1995 BASt developed a design for, but did not build, a multi-wheel system, similar to that built in Australia in 1960 (Figure 20.11) and the later, and larger, Texan system (Figure 20.21).

Extending the approach illustrated in Figure 20.20, the University of New South Wales developed a system with multiple plates loaded in a sequence in an attempt to simulate the horizontal movement of a wheel (Figure 20.22). Its main purpose was to test segmental concrete block pavements (Chapter 10).

Another outdoor facility is the two test roads owned and operated by the Public Works Research Institute (PWRI) in Japan, which was commissioned in 1970 (Figure 20.23). The system is unique in that the test vehicles are driverless, with their passage automatically controlled as they travel along the test loops. The test program, which is largely government-funded, concentrates on the performance of asphalt pavements, by far the most common pavement type in Japan.

Other pavement testing facilities that have not been reviewed in this chapter but which are discussed elsewhere by Metcalf[887] are (or were) in Alabama, Amsterdam, Arizona, Brazil, Brno (Czech Republic), California, Canterbury

Figure 20.21 Texan pavement loading machine developed in 1995.[886]

(Christchurch, NZ), China, Costa Rica, Crowthorne (TRRL England – now de-commissioned), Florida, Indiana, Kansas, Kentucky, Louisiana, Mexico City, Italy (Nardo), Nevada, Nottingham (UK), Pennsylvania, Poland, Canada (Saskatchewan), South Korea and the US Army Corps of Engineers at Vicksburg, Mississippi, and Lyme, New Hampshire.

Currently there are about 40 operating pavement testing sites around the world and the associated technology is widely called accelerated pavement testing (APT).[889] The continuation of the existing and spread of new programs indicates positive outcomes, and one major review[890] found that

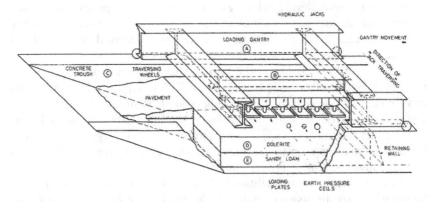

Figure 20.22 University of New South Wales test method for pavement loading.[885]

Figure 20.23 Pavement testing using driverless trucks, Japan.[888]

APT had served as a means of improving performance and economics of pavements by increasing the understanding of the factors that affect pavement performance by:

1. exploring a wide variety of pavement compositions and configurations,
2. simulating the mechanisms, conditions and processes due to loading and the environment,
3. testing and characterising materials,
4. analysing and understanding pavement response and performance, and
5. applying high wheel loads over a much shorter time than in the real world where individual vehicles are spatially separated. Not all those vehicles are trucks, and not all truck wheels are fully loaded.

The acquired knowledge has been widely applied and has enhanced innovation in pavements. The tests have aided the use of new materials and new methods of construction, maintenance and rehabilitation. Reviews suggest benefit/cost ratios of about 4.

Road-based testing was discussed at the beginning of the chapter. Applying controlled loads has removed a major concern with those earlier tests. A further advantage of these road-based test sites over the specific purpose-built sites or laboratory facilities was that it had become increasingly obvious that the performance of a pavement was very dependent on the way it was placed. The specific tests inevitably fell short of a real-world situation because specific sites and facilities were constructed very quickly and were far more closely supervised than would be encountered in normal construction practice. In addition, the test subgrade would contain no unknown imperfections. The specific tests would thus inevitably fall short of a real-world situation.

Chapter 21

Possible future pavements, with an emphasis on recycling

Every track, path and road has a surface; however, describing a way as "surfaced" implies that it has some form of practically useful pavement. The paved roads discussed in the earlier parts of this book were relatively rare, connecting major origins and destinations or within densely settled towns. Most lesser roads were unpaved. The advent of the motor vehicle at the beginning of the 20th century changed this circumstance completely. It is now possible to look back and see the subsequent growth of paved roads in the developed world following a classical logistic curve with the proportion of roads that are paved in many countries now approaching a 100% saturation level. In much of the developed world (but not in the authors' Australia) paved surfaces are now the norm rather than the exception.[891] This has many implications technically and in our perceptions of the place of pavements in our real world.

So where do pavements currently lie in their sequence of social and technical development? In the context of the various matters described earlier in this book, the current technology for building and maintaining major pavements is potentially very effective. There is no reason why a modern pavement should not deliver its intended system-wide outcomes. The choice of pavement type is usually between asphalt and concrete and is made on economic and political grounds as both can perform adequately. In most current situations, asphalt is the preferred option. This is often because bitumen, which is otherwise a waste-product of oil refining, can be priced to ensure that it is a slightly cheaper alternative than cement.

On minor roads, which are far more extensive than major roads, the cost of elaborate machinery and imported materials for pavement making and maintaining is usually prohibitive. Local materials are used and increased attention is paid to roadside drainage. The surfaces may be strengthened by a layer of compacted broken stone brought from local quarries. As a further improvement, the surface might be enhanced by covering it with a bituminous spray or a thin layer of asphalt (Chapter 14).

Three major differences over the last century relate to the traffic that the pavement must carry.

1. Traffic on major roads has always been heavy. However, the speed of modern traffic is an order of magnitude faster than before the 20th century. This allows more traffic to use a given length of pavement and means that the number of wheel loads on the pavement over time is much greater.
2. The widespread use of pneumatic tyres means that the local pressures on the pavement surface cannot exceed the pressure in the tyres. Improved vehicle suspensions have a similar softening effect.
3. Wheel loads are now much better controlled. A single overloaded wheel can destroy a pavement but this is now a far less common occurrence due to standardised truck and tyre design (Figure 21.1) and to devices for detecting overloaded trucks, when either stationary or in motion (Figure 21.2). Weigh-in-motion devices can be embedded in pavements or culverts and their output not only reduces overloading but also provides invaluable input for the design of future pavements. Another control on overloaded vehicles comes from the rise of fleet logistics and freight-forwarding which means that many trucks are loaded by quality-managed parties other than the truck operator.

Figure 21.1 An Australian road-train truck capable of legally carrying a load of over 100 tonne.[892]

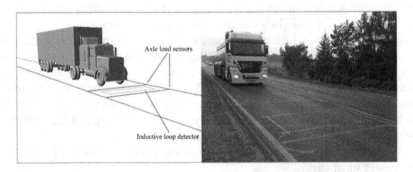

Figure 21.2 Weigh-in-motion in Poland.[893]

The downside of the increased frequency of high (but legal) wheel loads is that progressive failures such as rutting and cracking occur earlier in the life of a pavement, with the pavements consequently needing rehabilitation much earlier than would previously been expected. This has had problematic consequences for regions which have selected their pavement type from a catalogue of formerly successful pavement configurations. A better approach might have been to select new pavement configurations using one of the methods of analysis and design discussed in Chapters 13–17. As discussed in those chapters, these methods are still far from perfect. It is therefore appropriate to revisit those imperfections in summary form.

First, with respect to subgrades, there is still no universal objective method of assessing subgrade suitability. Furthermore, there is still an unrealistic belief that any subgrade, no matter how poor, can effectively and economically be raised to an acceptable standard.

Second, with respect to the pavement itself, there are still serious shortcomings in the methods of analysis and design. This criticism does not apply to concrete pavements which have been able to draw upon methods well-developed for the structural use of concrete in bridges and buildings. However, asphalt pavement design in some regions still relies on a catalogue of past "successful" designs, with occasional failures tolerated and success often seen through economically insensitive eyes. Other regions rely on algebraic methods based largely on often-unconvincing statistical analyses of tests on dedicated pavements. The common models of internal pavement behaviour relate pavement performance to pavement properties via relationships which may have little phenomenological basis. There often is no conceptual guidance as to why a change in an input should cause the specific change in a relevant output.

While this criticism of many asphalt pavement design methods might seem a little harsh, it arises from some very specific factors which mainly relate to the bituminous component of asphalt. Unlike the cementitious mortar used in concrete, the properties of bitumen vary continuously with applied stress, with temperature and with time. There is no simple way for the variations to be represented in an algebraic model. A criticism of many of the current asphalt design methods is that it would have been better if statistical power had been applied to a valid phenomenological model, rather than to creating an algebraic model that mimics the suggestions of a statistical analysis of the input and output variables.

Another problem with asphalt paving is that the bitumen used as the internal glue is supplied by oil refineries as a by-product of their refining process. Its properties depend on the oil being refined at that time and for a specific fuel-related sales purpose. Despite admirable past efforts by Shell, there is no commercial incentive for the refinery to do more than the simplest of tests, none of which relate usefully to the field performance of the bitumen. Furthermore, the supply of bitumen can be affected by events such as wars and embargoes, which may force customers to seek alternative sources, leading to them using bitumens with which they may not be very familiar.

Similarly, it should also be noted that the pavement layer of "compacted" broken stones is also very difficult to describe in a phenomenological way.

Looking forward, it is likely that future asphalt design methods will take advantage of more phenomenologically appropriate procedures. From a sustainability context, both asphalt and concrete pavements require major energy inputs and their cost will be sensitive to changes in energy costs. Although cement-making is energy intensive, the materials that it uses are relatively common. On the other hand, the availability and cost of bitumen depends directly on the availability and cost of petroleum and so the future of asphalt may be less certain than the future of concrete. In this context, a number of groups are looking at the use of organic resins as a future alternative to bitumen.

McAdam was a strong advocate of recycling the stones found in any pavement he was replacing or constructing (Chapter 9). Recycling is now a common part of a general sustainability assessment of a new pavement, recognising its potential to provide a significant reduction in the energy, water and materials 'footprint' caused by the construction and operation of a pavement. Thus, recycled materials are increasingly being used in pavement construction, mainly combined with bitumen to produce asphalt.

Materials recycled in pavements have included reclaimed concrete, asphalt, waste plastics, rubber, scrap tyres, glass, waste rock, crushed bricks, blast-furnace and smelter fly ash (Chapter 18), furnace slag, spent foundry sands and waste motor oil. Since about 1970, the recycling of aged pavements (Chapter 19), particularly surface layers, has become more common with the recycled material being used as a part of the rejuvenated pavement. In many places, asphalt is the community's most recycled material. One commentator claimed that American roads were the world's biggest asphalt quarry.[894]

A major limiting factor for the use of many recycled materials is their ability to bond adhesively with bitumen. For example, recycled glass does not absorb any of the bitumen and is also "hydrophilic" in that moisture can cause bitumen to strip from the glass. In 1997, the American national road agency (FHWA) stated that:[895]

> Waste glass that is crushed and screened can be used as a portion of fine aggregate in asphalt paving mixes. Satisfactory performance has been obtained from hot mix asphalt pavements incorporating 10 to 15 per cent crushed glass in wearing surface mixes. The term "glasphalt" has at times been used to describe these pavements. Higher blends, incorporating perhaps up to 25 per cent, could potentially be used in base or binder course mixes. Hot mix asphalt surface course pavements with more than 15 per cent waste glass may experience deterioration due to stripping of the bitumen from the waste glass.

Another significant problem is the cost of preprocessing waste/recycled materials together with the cost of transporting them from source to site. Flat and elongated particles are difficult to recycle as they can contribute to pavement ravelling, stripping, poor skid resistance, abnormally high tire wear and excessive glare.

It has been commonplace to locate utilities within a pavement or its associated subgrade, taking advantage of excellent access that the roadway provides to both the utility and any of its roadside consumers and producers. With the increased use of electric vehicles, it is possible that their power sources could be located within the pavement, perhaps transferring power by induction. Of course, since 1881, overhead electric cables have been used to transfer electrical energy to buses and trams via direct contact.[896]

In this respect, an interesting aspect of asphalt pavements is that their relatively black surface means that they could function as a form of black body absorbing solar energy and feeding some energy into a broader energy network. Some have considered recovering the energy that passing vehicles use in elastically deforming the pavement. Recent innovations have included heat-insulating pavements (solar heat-blocking pavements) aimed at improving the heat environment, particularly in urban areas, and water-retaining pavements to manage and collect rainwater.

Earlier chapters have emphasised the need to keep a pavement surface watertight to prevent water from damaging the pavement and the underlying basecourses and subgrade. However, impervious surfaces can cause significant undesirable changes to the hydrologic cycle in urban areas, including increased peak water flows and drainage runoff and reduced groundwater recharge. If pavement damage due to water can be prevented by using appropriate materials and configurations, there can be ecological advantages in having a permeable pavement. Such pavements can be inherently porous or discontinuously porous due to porous joints in interlocking block pavements. A related effect is to allow some plants to grow in the longitudinal drains beside a pavement, thus slowing water flow into downstream areas. A good example of this approach can be seen in Pinehurst, Seattle, website.[897]

Chapter 22

How pavements could be funded and operated in future years

A pavement will be part of a broader transport system and the benefits and costs of such a system will be part of a study of transport economics[898] and beyond the scope of this book. In our specific context, a pavement should fulfil its intended purposes and maximise the difference between:

a. the positive benefits it delivers to road users and the adjacent community, and
b. its construction, maintenance, operating and replacement costs and any costs it might impose on the broader community.

The book has explored the historical role of pavements. The first pavements arose to satisfy the local needs of early human communities and were never an end in themselves. This indirect motivation has continued over time and explains why pavements have never been a primary focus of human attention – they have always been just one of the means by which some broader set of objectives could be met. Thus, pavements were commonly underfunded and poorly maintained and rarely attracted the most influential and talented people. The one notable exception to these circumstances was the Roman road system, as discussed in Chapter 6. For the Romans, the economic and administrative benefits of building and maintaining a good system of paved roads were very apparent.

Elsewhere and at other times, the construction and maintenance of paved roads was a local matter and received little priority as most of the beneficiaries of the paved roads lived outside the adjoining area. Taxation systems were usually too primitive to rectify this disconnect. The first significant change occurred in medieval France where a strong national administration based in Paris recognised the need for a sound system of paved roads focussed on the national capital. The French not only devoted significant resources to the task but also trained and appointed military-style officers to implement and manage the system. Then, in the 19th century, Britain's economic dominance placed new emphasis on the importance of improved land communications, initially via strong pressures from the Post Office and then from the rise of independent and privately funded toll road companies. These companies had an economic imperative to provide good and cost-effective paved roads. This need gave rise to the paving contributions of John Loudon McAdam. His story is discussed in Chapter 9 and he was, in our view, the greatest individual contributor to pavements over the course of human history. Nevertheless, his new methods were often viewed with scepticism and it took over a hundred years for many of the misinterpretations of his work to be put to rest.

The beginning of the 20th century saw the rapid development and widespread adoption of self-powered vehicles using internal-combustion engines fuelled with petroleum distillates. For pavements, there were three major consequences. First, there was a massive increase in the demand for paved roads.

The second consequence was that governments began using a tax on the sale of the fuel for the new vehicle to fund the construction and maintenance of pavements. The prime example of this approach occurred in the mid-1950s when the US government began using fuel-tax to construct its massive "Interstate" freeway system. However, despite the development of the pavement management systems discussed in Chapter 19, it still is quite rare for governments to allocate adequate fuel-tax revenue to the provision and maintenance of pavements.

The third consequence was that a serendipitous by-product of producing petroleum distillates was a thick, sticky material called bitumen. Most readers, if asked to name the most common paving type, would correctly nominate asphalt which is a mixture of bitumen, sand and stones. Asphalt is found naturally and was probably first used for paving in the original Mesopotamian empires. Nevertheless, its use for practical paving did not occur until early in the 19th century and then largely by accident. As discussed in Chapter 13, for at least the next century its use was befuddled by technical misunderstandings and deliberate commercial obfuscations. These were partly resolved by the end of World War I, but another quarter century was to pass before a good technical basis was developed for the use of asphalt pavements. A core feature of this development was the realisation that excellent asphalt could be produced by the controlled mixing of bitumen, sand and local broken stone. As discussed in Chapter 14, bitumen is also the basis for the most cost-effective way of enhancing the characteristics of low-cost non-urban roads. Similarly, it was also found that good pavements could be produced by using limestone-based cement rather than bitumen.

The two major technical issues encountered with the provision and operation of pavements have always been the managing of water and of heavily laden trucks. We saw in Chapter 3 how many of the early methods for managing water were based more on religious beliefs than on common sense. The rise of rational approaches in the 18th century led to people like Trésaguet, McAdam and Telford bringing sanity to the fore in managing the effects of water on pavements.

The second issue was measuring and then monitoring truck wheel loads. These are not simple tasks and as traffic increased from the 17th to the 19th century, there was an extraordinary proliferation of geometric limits on the wheel size in a futile attempt to indirectly control wheel loads (Chapter 8). Heavy wheel loads can destroy a pavement in a single pass and, in the absence of effective regulation, many pavements were saved solely by the inability of truck operators to move their overloaded vehicles when their wheels quickly sank into the underlying mud. The problem was largely resolved by the introduction of self-regulating pneumatic tyres early in the 20th century (Chapter 19).

A major new pavement impact occurred with the use of internal-combustion trucks during World War I. When used on civilian roads during and after the War, they rapidly destroyed pavements due, not to the magnitude of the individual wheel loads, but to the frequency with which those loads occurred and to the high contact stresses between the pavement and their solid tyres. Pavements failed quickly due to fatigue, and some decades were to pass before the problem was understood and rationally managed (Chapter 16). Again, pneumatic tyres were a major part of the solution. World War II had a more positive impact on pavement technology as new methods developed for quickly and cheaply building airfield runaways soon flowed through to civilian pavement practice.

As we conclude this history, asphalt is by far the most dominant paving material. Its production and placement are now quite sophisticated. It still has two key sets of problems. First, it relies totally on bitumen, a material that is a by-product of other industrial processes and whose availability depends on the availability of petroleum with all the associated geological, economic and geo-political influences on that availability. Second, its properties depend on its source, construction practice, time in service, traffic usage and temperature and so its future performance will always be far less predictable than would be the case with other more conventional structural materials. Concrete is the only practical alternative and can compete with asphalt in various situations (Chapter 17).

In many respects, the management issues discussed earlier nowadays overwhelm the technical issues raised in the preceding paragraph. The construction of new pavements and the maintenance and replacement of worn-out pavements require significant planning, funding and ongoing management. The pavement users and beneficiaries are rarely directly involved in decisions on any of these three matters. There are many current examples of where this situation is still being poorly addressed, and very few examples of it being properly addressed. Perhaps the best examples are some of the current toll roads where the application of sound business practices appears to be avoiding many of the problems associated with the more common "public" pavements which are funded indirectly and managed via political processes with their associated short-term electoral and financial cycles.

Given the various factors discussed above, it is easy to be critical of the paved roads which various human societies have created and used over six millennia. However, it is important to realise that these ways were built and maintained with great human efforts, in the face of many challenges and in far less than ideal circumstances.

From the first footpath, paved ways have increasingly pervaded all aspects of the manner in which our land-based world operates. Although they are now a major part of the fabric of our daily existence, their extraordinary contribution has been but rarely noted, let alone lauded. Hopefully this book will help to change the way in which our pavements are viewed.

Finally, an example to show that the end is also the beginning. Highway Route 339 runs around the Tsugaru Peninsula at the northern end of the island of Honshū in the Aomori Prefecture in Japan. It was upgraded to a National Road in 1974. It has a unique finish at the far northern end of its 108 km journey as it includes a 338 m length of pavement with 362 steps provided to help students travel safely to two schools located on the slope. The Aomori Prefecture made improvements to the steps in 1993. The "National Highway with steps" is now a popular tourist attraction (Figure 22.1).

The situation perhaps recalls TS Elliot's observation in his *Four Quartets* that it is only when we arrive back where we started that we know a place for the first time.

Figure 22.1 Route 339 in Japan, a National Highway with steps.[899]

Timeline

The following table presents a timetable of events, in chronological order, in the development of the world's road pavements. Some of the dates are approximate. As discussed in Chapter 1, this table does not include the natural development of pathways in the earliest history of man.

Approx. Date	Event
2600 BC to 1100 BC	The Minoans in Crete develop relatively sophisticated road-building methods which probably influenced later Etruscan and then Roman road builders.
2500 BC	Road in Egypt's Old Kingdom constructed to enable basaltic stone to be brought to Giza for building pyramids and temples. About 12 km of that road – now known as the Lake Moeris Quarry Road – still exists.
2000 BC	Pavements of bricks mortared with bitumen constructed in Iraq and Pakistan.
	Wheeled vehicles common in the Middle East and parts of south-eastern Europe.
1600 BC	Stone-surfaced Minoan road built near Knossos.
Pre-1500 BC	Various examples of corduroy roads built in Novograd in Russia, near Prague, around "lake" villages in Switzerland, in the Pamgola swamps in Hungary, between Magdeberg and Scherwerin in central Germany and elsewhere in the flat and swampy loess and peat lands of northern Europe, southern Holland, and Glastonbury.
1200 BC	Earliest paved street constructed by the Hittites in Boğazkale in central Turkey.
1170 BC	Stone-paved ceremonial route built between two temples at the holy city of Bubastis, about 40 km north of Cairo. Herodotus described it as about 100 m wide and bordered by trees so tall that "they seem to touch the sky".
700 BC	Stone pavement constructed at Nergal Gate, Nineveh, near Mosul.
	Brick pavements surfaced with limestone slabs constructed in Iraq and Babylon.

(*Continued*)

Approx. Date	Event
600 BC	Processional Way built by Nebuchadnezzar in Babylon. It was almost 2 km long and paved with limestone flagstones placed on three horizontal layers of burnt bricks, all mortared together by lime and bitumen.
400 BC	Cement being used by the Greeks and Etruscans (and then the Romans).
	Etruscans in central Italy were paving the streets of their towns; four paved roads from this era found in Bologna.
	The streets of Vetulonia, about 200 km north of Rome, are paved in stone by the Romans.
334 BC	First arterial road built by the Romans: Via Latina from the Roman gate, now known as Porta San Sebastiano, to Capua.
	Via Latina replaced by much straighter Via Appia (Appian Way) in 312 BC. By 264 BC the road had been progressively extended a further 400 km across the Italian peninsula to the Adriatic port of Brindisi. It remains in restricted use today.
320 BC	Earliest recorded toll system introduced in India.
221 BC	Qin First Emperor in China mandates a standard wheel spacing to ensure that carts and carriages could use existing road ruts.
200 BC	Chinese road builders use metal rammers to compact roads.
	By the 19th century European rammers weighed about 20 kg and had iron bases.
7th century	Japanese road system begins to develop: about 6.5 Mm of arterial routes brought into operation over the following century.
850	Flagstone paving used in Córdoba in Spain during the reign of Caliph Abdorrahman II. It had previously been used by the Romans, particularly in Rome.
1184	The stench of Parisian streets so offends King Phillip II that he orders all of its streets to be paved so that it would be possible to remove the daily accumulation of excremental muck that was deposited by transport animals and by adjoining citizens who dumped all their personal and domestic waste and dead domestic animals into the streets.
1280	Judge Phillipe Beaumanoir defines Roman roads in French law as the 5th and greatest class of the nation's roads.
1282	First recorded use of the word *pavior* (stonemasons working as road builders)
1285	Edward I notes that the road through Dunstable was "so broken up and deep, that dangerous injuries continually threatened travellers".
1315	Edward II issues a writ on the Mayor of London requiring London pavements to be "repaired, cleansed and freed of vagrant pigs".
1415	Cubical blocks about 150 mm in size used in Paris.
1474	Maud Heath uses her life savings to build a causeway, partly through the bog between her house and the Chippenham market where she sells her eggs. It is still in service and still carries her name.

(Continued)

Approx. Date	Event
1353	Edward III issues an order requiring paving, with the work funded by adjoining land-owners and by all vehicles and laden beasts bringing merchandise into London. He had authorised tolls on roads near Temple Bar in London and in Dublin in 1346.
1499	An unsuspecting traveller on horseback falls into a huge pit in the road after a miller removed the clay to repair his mill. Both the traveller and his horse drown. The miller is charged with murder but is acquitted as it was considered that he had had no malicious intent as the highway was the best place to find the clay he needed.
1502	An ordinance prohibits Parisians from tipping waste out of windows; this was gradually relaxed to a requirement that they shout *gardez l'eau* before tossing.
1532	English Statute under Henry VIII requires London roads to be made "using cobblestones brought from the seashore".
1547	Penalties for poor paving in France had increased from fines, to flogging, imprisonment and hanging.
1555	The first device for weighing carts and wagons is introduced in Dublin – these lever-arm device (weighbridges) were used more to collect tolls based on load carried rather than to prevent pavement damage.
1562	Act gives English parish road "supervisors" the authority to divert any "water-course or spring of water" into a ditch adjoining the road.
1585	Guido Toglietta in Italy writes a treatise on pavements in which he suggests that lighter pavements than those used by the Romans would be feasible and effective.
1594	Henry IV of France appoints a senior official (the Grand Voyer de France) to manage the national road system. The position is abolished in 1627.
1595	Sir Walter Raleigh uses bitumen from the Trinidad Asphalt Lake to caulk his wooden ships. Its bitumen is being continuously replenished from underground sources. It is probably the world's largest bitumen deposit.
1610	First book on roads written in English produced by Thomas Proctor.
1622	A detailed technical account of Roman roads produced by Nicholas Bergier. It led to a lavish but technically unsuccessful revival of Roman road-making practice in France and in a number of German States.
1661	Tolling given legal status in the UK. The first toll house opened on the road at Wadesmill, about 35 km north of London. Evasion is very high and an Act in 1695 allows barriers (toll gates) to be erected.
1665	Bishop of Münster, during a war between Germany and Holland, builds a 6 km long road though Bortanger Moor by supplementing brushes with pieces of houses destroyed in nearby villages.
1690	Chancellor Cooper writes to his wife attributing the long legs of Sussex women to the constant pulling of their feet out of the mud as they travelled, which would *tend to strengthen the muscles and lengthen the bones.*

(Continued)

Approx. Date	Event
1693	Henri Gautier, who had originally trained as a Doctor of Medicine, is appointed the national Inspector General of Bridges and Roads in France. In the same year he publishes a book on pavement construction.
1706	An Act enacted in Britain gives private trustees the power to collect tolls on lengths of roads that they have to (at least in part) construct and fully maintain. They are called "turnpikes" as their gates resemble military weapons called pikes.
1711	Eirini d'Eyrinis (or Eirinus), a native of Russia, a professor of Greek and a doctor of medicine, obtains a 25-year concession from the owner of the Val de Travers bitumen deposits, the King of Prussia, to produce bitumen-based medicinal products, mastics for waterproofing and for caulking ships, and portable fuels. He publishes *La Nosourotechnie* in 1712, the first publication on bitumen in the world.
1717	"Blind" John Metcalf born in Knaresborough in the West Riding of Yorkshire. He was blinded by smallpox when he was 6 years of age and received no further formal education other than music lessons.
1725	Daniel Defoe writes: *The inland trade of England has been greatly obstructed by the exceeding badness of the roads.*
1726–1727	Kensington, Fulham and Chelsea Turnpike Trust purchases 73,000 bavins (bundles or faggots).
1745	British government almost overthrown on account of the bad state of British roads.
1747	École Nationale des Ponts et Chaussées established to train road engineers in France
	First recorded use of bituminous paving in Europe: a mixture of bitumen, gravel and rubble used to produce an asphalt carriageway in the Dublin home of the Earl of Leinster.
1756	John Loudon McAdam, Scottish civil engineer and road-builder and the inventor of "macadamisation", born.
1767	Droitwich Toll Road Trust in Worcestershire issues a notice directing that *Persons must not take away manure from the side of the road as it is wanted for filling up the holloways.*
1775	A-R-J Turgot, Louis XVI's Finance Minister, appoints Pierre-Marie Jerome Trésaguet as Engineer-General of the Corps des Ponts et Chaussées. Trésaguet believes that the structure used by the Romans is "unnecessarily massive".
1796	James Parker grinds and burns lime-clay nodules he had found on the Isle of Sheppey in Kent and uses the material to produce a practical cement.
1803	Scottish stonemason and bridge-builder, Thomas Telford, receives a major and well-funded government commission to provide roads as a consequence of rebellious Jacobite troubles in the largely inaccessible Scottish Highlands.

(Continued)

Approx. Date	Event
1806	Napoleon introduces a tax on salt to pay for road works but the funds are diverted to other uses. After the 1830 Revolution, a law in 1836 provides reliable revenue for roads based on a tax on all able-bodied males.
1810	"Blind" John Metcalf dies at the age of 93. He had 90 great-grandchildren.
1816	The first full-scale pavement test is conducted by a committee of the Dublin Society under the leadership of Richard Edgeworth. The test shows that a broken stone pavement requires much lower haulage forces than pre-macadam alternatives.
1817	Phillip Clay in England develops a horse-drawn roller which could apply a load of up to 20 ton.
1820	John McAdam's name given to the term *macadam*, used to describe courses of compacted broken stones working together due to natural interlock between the stones.
	Wooden blocks used by Gourieff (a Member of the Russian Academy) on Great Sea Street and Million Street in St Petersburg in Russia. St Petersburg hexagonal blocks exported for use in street paving in Manchester in 1838.
1820s	English Bridge at Shrewsbury uses white stones as line-markers to separate oncoming traffic.
	Telford's assistant, John Macneill, develops and patents embedding 25 mm cubes of iron in the pavement surface at 100 mm centres. No positive outcome has been reported.
1822	First McAdam pavement placed in Australia between Prospect and Richmond in NSW.
1824	The firm of Pillot & Eyquem uses Seyssel powdered rock asphalt to produce asphalt blocks to pave parts of the Champs Elysée. The trial is extended in 1837 when asphalt blocks are used on the Place de la Concorde.
	Joseph Aspidin, a bricklayer from Wakefield in Yorkshire, invents modern cement by modifying the clay content and the burning temperature. He names it Portland cement as it resembles the colour of the limestone quarried on the Isle of Portland in Dorset.
1825	Thomas Telford uses Portland cement to build a 5.4 m wide and 150 mm thick concrete slab. It is employed for 2.5 km of pavement near Highgate Arch in London, as a test for use on his newly commissioned Holyhead Road. The test is successful.
1829	The US State of Georgia repeals its statute labour legislation and replaces it with a $70,000 allocation to purchase slaves to do the (road) work. (The ancient Greeks had similarly used publicly owned slaves to build and maintain public roads.)
	John Macneill develops a *road indicator*, the first device to specifically measure road condition. It uses a dynamometer based on existing weighing machine technology using weights and lever arms to assess the force required to pull a vehicle at a known speed along a road.

(Continued)

Approx. Date	Event
1835	Darcy Boulton develops Yonge Street in Toronto using a version of the corduroy road method that came to be called a *farmers' railroad*; it is subsequently widely marketed in North America and Australia.
1836	Compressive strength testing of concrete cylinders is first used (in Germany).
1840	First tar penetration macadam pavement is placed on a 3 km length of Lincoln Road outside of Nottingham; it performs poorly.
	Polonceau introduces the first heavy steam-powered rollers into France.
	Rubber slabs are tested as a pavement surfacing on the Admiralty Courtyard in London.
1846	First plank road is built between Syracuse and Oneida Lake in New York State.
1850	The French begin using steam-powered rollers during trials of asphalt on the Avenue de Marigny in Paris.
1850–1871	Wood block pavements impregnated with tar & creosote catch fire in San Francisco (1850), New York (1870) and Chicago (1871).
1850s	Steam-powered stone-breaking machines first produced by Eli Blake in Connecticut.
	Stone-crushers built to an 1858 Blake patent were being used in Britain by 1862.
1853	The earliest known mechanical device to measure longitudinal profile (Orograph) is produced for the US Army. It uses a long trough of mercury to provide a horizontal datum and then draws the profile on a paper roll.
1855	Louis W Clark in Kolkata (Calcutta), dismayed by the problems he is having with road rollers drawn by bullocks, asks W Bathe in Birmingham to produce a steam-powered roller. It is sent to Kolkata in 1864.
1859	First British Ambassador to Japan writes glowingly of Japan's Tokaido Road, which was developed during the 17th and 18th centuries (today it links Tokyo and Kyoto).
1865	Paris municipality awards a 6-year contract to the Paris Steam Road Rolling Company to provide seven steam rollers for the continuous use of the municipality.
	Thomas Aveling builds his first road roller using his traction engine production line in Rochester, Kent. Initially used in the maintenance of macadam roads, crushing the surficial stones and smoothing the surface, steam-powered rolling soon replaces the last two manual stages in asphalt placement and greatly increases the usefulness of asphalt as a pavement surfacing.
	Joseph Mitchell, who had trained with Telford, introduces the cement penetration method in which a cement-sand mortar spread on top of a macadam pavement layer is swept and pushed into the interstices in the macadam course.

(Continued)

Approx. Date	Event
1869	French authorities introduce the Deval test for measuring the abrasion resistance of broken stones.
	The micro-Deval test developed in France in the 1960s is a modern version of this test.
	A Seyssel mastic asphalt is placed over granite block paving in Fifth Avenue in New York. The pavement begins to fail within three weeks and has to be totally removed. The cause is attributed to contractors with inadequate knowledge and experience in using the new product.
	First English trial of powdered rock asphalt (from Val de Travers quarry) in London on Threadneedle Street near Finch Lane.
1870	Over 75% of Parisian streets are now paved with setts.
	The use of a powdered rock asphalt in Paris peaks due to the demands of the Franco-Prussian War. Its use then declines due to poor experiences with maintenance and slippery surfaces.
1871	Edward de Smedt, a Belgian chemist, who had emigrated to work at Columbia University, mixes heated Trinidad bitumen with sand and powdered limestone to produce a mastic which is used successfully on William Street outside City Hall in Newark, New Jersey.
	First American patent for asphalt taken out by Nathan Abbott in New York. It is based on a tar-based asphalt pavement he had placed in Washington in 1868.
1874	The belief that the vapours rising from damp wood block paving were a carrier of fatal diseases such as typhoid, dysentery and malaria – and their failure in service after 4 years – is said to have been the primary cause for the City of Washington losing home rule and becoming a federal District.
1876	Edward de Smedt is appointed to a new position of Inspector of Asphalts and Cements within the ambit of the US Army Corps of Engineers.
1877	Equipment with scraping blades held by mechanisms that permits their operating angles to be adjusted ("American Champion") patented by Samuel Pennock in Pennsylvania.
	Leaning wheel versions introduced by Joseph Adams in Indianapolis in 1885.
1879	Deacon notes that *foremen accustomed to the work can judge whether it (tar) is too brittle by biting it.*
	In 1888 Bowen of Barber Asphalt replaces oral test with a test in which a "No. 2 sewing needle" is poked into the bitumen. It becomes the penetration test.
1880	Adrien Mountain conducts a trial of eight different types of wood blocks in Sydney; the hardwood blocks perform very well. As a result, softwood blocks in European cities are often replaced by hardwood (eucalyptus) blocks imported from Australia.
	First well-publicised trial of *penetration macadam* is conducted near Bordeaux in France.

(Continued)

Approx. Date	Event
1881	Overhead electric cables are first used in the USA to transfer electrical energy to buses and trams via direct contact.
1885	Joseph Valentin Boussinesq provides a solution for stresses below the surface of a layer of elastic material due to a point load on its surface.
1886	Adrien Mountain – pioneer of wood block paving in Australia – is appointed City Engineer of Melbourne.
1890	The first edition of American magazine, *Paving and Municipal Engineering*, includes an anonymous article on its first page headed *The Accursed Cobble-Stone*.
1892	The square outside the Main Street courthouse in Bellefontaine in Ohio is paved by placing concrete directly on the subgrade. Parts of this pavement were still in service in 1967.
1896	*Hot mix asphalt* is first used at the Barber Asphalt paving plant in Long Island, New York City.
1900	Road-builders Walter Gillette and John Fitzgerald in California observe how well a flock of sheep compact their unfinished oiled pavement. It leads to the development of the sheepsfoot roller for compacting clayey and cohesive soils. The first version was iron spikes hammered into a log. Earlier, 18th-century canal builders had used herds of cattle to tamp into place the layers of clay used to waterproof the surfaces of their canals.
	John Brown, President of the Irish Roads Improvement Association, produces his Viagraph, a 4 m long straight-edge with a wheel at one end. It records the longitudinal profile of a pavement on a paper roll.
1901	Purnell Hooley observes that a barrel of tar, which accidentally bursts over a layer of blast furnace slag on a road near the Denby Iron Works in Derbyshire, produces an admirable road pavement. After some further investigation, including adding pitch, cement and resin to the tar, Hooley produces a useful pavement and patents his product, which he names *Tarmac*.
	The first asphalt production facility to contain most of the basic components of a modern asphalt plant is assembled by the Warren Brothers. The seven brothers had commenced business in 1890 in East Cambridge, Massachusetts, as an asphalt roofing firm.
1902	Ernest Guglielminetti develops a method for spraying tar on roads, initially on the Grand Corniche road outside the Casino in Monte Carlo. He came to be known in France as the Tar Doctor (le docteur Goudron).
1905	Rudolf Diesel, an avid long-distance motorist, describes the road dust generated by his vehicle when travelling along a valley in Italy as causing "a great public nuisance".
	Pressure from users of the new internal combustion vehicles leads to the US Office of Public Road Inquiries being renamed the Office of Public Roads. Logan Page, a geologist from Massachusetts who had spent time with the French Laboratoire des Ponts et Chausées, is appointed as its Director.
	The first planned artificial loading device is a trafficking machine designed by Irish engineer RJ Kirwan. It appears it was never built.

(*Continued*)

Approx. Date	Event
1906	Modern stabilisation applications begin in the American south, followed by the granting of American patents in 1910.
1907	Some 50 million wood blocks are in use as street paving in Sydney and Melbourne.
1908	At the first PIARC Congress in Paris, the third "question" on its agenda was, "How to reduce wear and dust?".
	Clifford Richardson emphasises that the role of a concrete basecourse is to support the foundation that carries the traffic loads and that the role of the asphalt is to protect that basecourse from wear and disintegration.
1909	A circular pavement test track is built in Detroit by JC McCabe, a boiler inspector in the city's Department of Public Works. It is called the *paving determinator*. One of its circulating wheels carries five simulated hooves and horseshoes while the other is a steel-rimmed cart-wheel to simulate the then current vehicular traffic.
1910	An expert from the American Office of Public Roads optimistically suggests that the only solution to the dust problem is to treat new and old pavements with "chemical substances".
1911	Swedish physicist, Albert Atterberg, devises a test for determining the moisture contents at which a particular soil becomes plastic (its *plastic limit*) and then effectively liquid (*liquid limit*), with any fine material in suspension in the water (Atterberg Limits).
	First use of painted white centrelines on bridges and bends – in Wayne County in Michigan.
	Experience with British trial road sections leads to the suggestion that similar information could be obtained more rapidly *by means of a machine which should apply to samples of various surfacing material the same sort of stresses as took place on the road*.
1912–1923	State of Illinois conducts the Bates Test Road. The results lead to the use of a longitudinal centre joint to eliminate longitudinal cracking. It also helps in the development of simple analytical methods to predict pavement stresses, strains and deflections.
1913	The first major concrete highway – a 40 km long pavement near Pine Bluff, Arkansas – is constructed in the USA.
	A 10 km controlled-access highway called Avus (*Automobil-Verkehrs – und Übungsstraße* – automobile traffic and practice road) is constructed in Berlin. It is originally built as a racetrack but is also used for testing vehicles and road surfaces.
1914	American Association of State Highway Officials (AASHO) is created by 18 of the American States. It becomes AASHTO in 1973 – the T stands for Transportation. It was then, and still is, the body representing all American State and Federal road agencies.
1918	The US Bureau of Public Roads (now Federal Highway Administration (FHWA)) launches a series of road tests at the Department of Agriculture's Experimental Farm in Arlington, Virginia, to study the factors seen as important in pavement design.

(Continued)

Approx. Date	Event
World War I	The use of internal-combustion trucks on civilian roads during and after World War I rapidly destroys pavements due, not to the magnitude of the individual wheel loads, but to the frequency with which those loads occur and to the high contact stresses between the pavement and their solid tyres.
1919	Thomas MacDonald, Director of the Bureau of Public Roads, sends a fleet of 79 trucks and about 300 soldiers (including Lt Col. Dwight D Eisenhower) across America from Washington to San Francisco. The trip takes 56 days and highlights the need for good pavements as much as General John J Pershing's desire to demonstrate the military effectiveness of trucks. After he becomes President, Eisenhower begins the construction of the American Interstate freeway system.
	Self-powered graders first produced by Richard Russell in Minnesota.
	Asphalt Association (later Asphalt Institute) created by US asphalt producers.
1920	Competitive American road-makers are specifying 102 different grades of bitumen. Federal action reduces this to nine by 1923.
	State of Illinois commences operation of the Bates Test Road to find a cost-effective way of meeting local demands for better roads. The 3 km long test pavement runs from Bates to Farmingdale Road on what is now the old US54. There are 71 test sections constructed using different materials and to different designs. In 1926 highway officials estimate via the Associated Press that the pavement tests have saved the State $9 million in construction costs.
1921	An elliptical test road sponsored by Columbia Steel Company is constructed in Pittsburgh, California. It is intended to determine the benefits of using reinforcing steel in concrete pavements.
1920s	American Bureau of Public Roads "trilinear" or "triangular" soil classification scheme developed (first published in *Public Roads* in 1924).
	Prevost Hubbard and Frederick Field introduce a test to detect the rutting propensity of asphalt, particularly sheet asphalt. The test leads to the widely used Hubbard Field method of asphalt mix design. Its use persists in some areas of the USA until the 1970s.
	The use of machines for spreading and finishing asphalt is introduced in the USA, beginning with a Barber-Greene machine.
1923	Arthur Talbot examines the theoretical packing of spheres and produces a widely-used grading equation to aid in the design of concrete mixes.
	First device to measure road roughness (Lockwood Roughness Integrator) is developed in the USA. It utilises the vertical motion of a moving vehicle, initially by measuring the vertical accelerations of the rear axle of the vehicle. It is simplified in a 1925 version (Relative Roughness Determinator) which measures the relative movement between the vehicle body and the front axle.
1924–1928	The period in the USA during which asphalting changes from *hand labour and animal power* to *mechanical equipment and gasoline power*.

(Continued)

Approx. Date	Event
1926	Harold Westergaard publishes equations for determining the stresses in, and deflections of, loaded concrete pavements. He introduces a term called *modulus of subgrade reaction* to account for the stiffness of the subgrade.
	The use of solid rubber tyres on trucks is phased out.
	Full-scale studies addressing surface friction/skidding commence in the UK at the National Physical Laboratory. Work is transferred to the newly-formed Road Research Laboratory at Harmondsworth in the early 1930s. It continues at the Transport and Road Research Laboratory at Crowthorne.
1927	A series of pavement failures in California highlights the need to know subgrade properties before selecting asphalt pavement thicknesses.
	Line-marking machines introduced in England.
	The use of automatic control of screeds to level finished asphalt is introduced in Orange County, California.
	Ralph Proctor of the Los Angeles Bureau of Waterworks shows that there is a maximum density that can be achieved with a given moisture content and level of compaction. The test to determine these values is now known as the *Proctor test*.
1928–1929	119 grades of cutback bitumen are in use in the USA. Federal action reduces the number to six.
1929	An oval test track developed in co-operation with Dutch engineers is operating in Bandung, Indonesia. It uses two tethered vehicles, a truck and a two-wheeled goods cart (*grobak*).
	California Division of Highways introduces the *California Bearing Ratio* (*CBR*) test to assess the structural usefulness of a subgrade.
late 1920s	To avoid roughness results being dependent on the characteristics of the vehicle, the US Bureau of Public Roads standardises a device called the Dana Automatic *Roughometer* (BPR Roughometer) which is mounted in a special standardised trailer towed behind a car being driven at about 30 km/h.
1930	First field use of cement stabilisation, in South Carolina in 1932. The method is formally codified by the American Portland Cement Association in 1935. It is used with great success to build airfields during World War II.
	Germany begins to build its autobahn system after Adolf Hitler comes to power. By the end of World War II it had constructed some 4,000 km of road. As steel for reinforcement was often unavailable, many of the pavements are plain concrete.
	The floating screed, which eliminates any need for side formwork, is introduced.
1932	FM Hanson develops a measurement-based method for the design and construction of spray and chip seals in New Zealand. His rational approach remains unchallenged for the next 50 years.
	The addition of natural rubber to improve bitumen ductility is initiated by Swiss engineers and first applied in Sheffield in England.

(Continued)

Approx. Date	Event
1933	Stone setts replaced in heavily-trafficked parts of London by cast-iron blocks.
1935	*Los Angeles Abrasion test* adopted in the USA, despite it being very noisy and sometimes giving a poor representation of how the stones would perform under traffic.
1936	Harry Heltzer, working for 3M in Minnesota, invents reflective glass beads that make painted line-markings far more effective.
1937	A standard textbook on tar somewhat blissfully comments that *All tar warts are not necessarily malignant but are characterised by relatively rapid growth…If treatment is applied in the early stages a complete cure is certain.*
	America's first paved tollway, the 270 km Pennsylvania Turnpike, is constructed using concrete for all of its pavements. By the end of the 1950s, asphalt dominates most paving markets.
1938	Bruce Marshall, Mississippi Department of Highways, develops a mix design process which is lighter and more portable than the Hubbard-Field device (*Marshall mix design method*). It is widely used for airfield construction during the Second World War. A further development of this approach was produced by Francis Hveem in California; it is popular in the USA in the 1950s.
1938–1940	First CBR-based pavement design chart published. It is the first time that pavement thickness is numerically linked to subgrade properties.
Late 1930s	Donald Burmister develops a solution for the stresses in a two-layer pavement and, in 1945, a three-layer pavement.
1942	Water-bound macadam is now only being used for very lightly-trafficked pavements in the USA.
1943	Western Association of State Highway Officials, a division of 11 western American States within AASHO, joins with BPR and the Highway Research Board to conduct the Western Association of State Highway Organizations (WASHO) Road Test to establish load limits for asphalt pavements.
Mid-20th century	The first practical device for measuring pavement surface deflection under a static load (*Benkelman Beam test*) is developed by Albert C Benkelman for use in the WASHO Road Test.
1951–1954	A group at the UK TRRL modifies an Izod pendulum device used for impact testing of metals to swing a standard piece of rubber across a pavement surface and measure the energy lost in the process. It is named the TRRL *Pendulum Tester*.
1952	First full-scale laboratory circular test track in continental European – in Iassy, Romania – commences operations.
	AASHO Road Test is conducted in Ottawa, Illinois, on six two-lane test road loops on the future alignment of IS 80. It results in the performance equations and nomographs used to develop a new pavement design guide, first issued in 1961, as the *AASHO Interim Guide for the Design of Rigid and Flexible Pavements.*
	The US government begins using fuel taxes to construct its massive "Interstate" freeway system.

(Continued)

Approx. Date	Event
mid-1950s	Ivan Mays of the Texas Highways Department develops a vehicle-based device – Mays Roadmeter (or Ride Meter) – which measures relative movements between a vehicle 's body and the rear axle and a more user-friendly data-logger held on a passenger's lap.
1958–1960	The concept of "pavement management" is introduced, largely in response to numerous unanticipated pavement failures on the US Interstate and Canadian Highway Systems.
1961	AASHO *Interim Guide for the Design of Rigid and Flexible Pavements* first issued.
1967	Public Works Research Institute (PWRI), Japan, commissions test roads. The system is unique in that the test vehicles are driverless, with their passage automatically controlled as they travel along the test loops.
mid-1960s	Stone mastic asphalt (SMA), also called stone-matrix asphalt, is developed in Germany; the first SMA pavements are placed near Kiel.
1968	Council for Scientific and Industrial Research in South Africa commissions the *Heavy Vehicle Simulator (HVS)* intended for use on in-service pavements. There are now about 18 units operating worldwide, and a large knowledge base has been developed with the data shared by HVS users.
1970	Ralph Haas and W Ron Hudson publish their book *Pavement Management Systems*. It is republished in 2005 with Lynne Cowe-Falls as a co-author.
1971	Researchers at the Franklin Institute in Philadelphia publish a design guide for porous asphalt pavements.
1972	The Californian ELSYM5 program is issued by the FHWA. A number of computer-based asphalt pavement design programs are now in use around the world. Finite element models (FEM) are later developed.
1977	*Accelerated Loading Facility (ALF)* – initially funded by all Australian state road agencies – commences operation; it had made its public debut on the forecourt of the Sydney Opera House in 1983. The facility has been operating continually since then; ALFs are also operating in the USA and China.
1984	The World Bank published details of the *International Roughness Index* (IRI). A key advantage is that it depends only on the longitudinal profile of the pavement surface and has no subjective component. Its *Highway Design and Maintenance Model* (HDM) is released the following year. Many enhancements have occurred since then.
1985	The importance of the statistical variation of pavement materials, their design, how they are constructed and maintained and how they are used, particularly in the context of the new processes of total quality management (TQM), becomes more widely recognised and understood.
1987	The Strategic Highway Research Program (SHRP), is established by the American Congress in 1987 as a 5-year, $150-million research program to improve the performance and durability of the US highways. One output is the *Superpave* method for pavement design and construction. It is designed to replace the Hveem and Marshall methods through the use of a Gyratory Compactor.

(Continued)

Approx. Date	Event
1991	Accelerated pavement trials of pavements with geotextile reinforced seals are conducted in Brewarrina, western New South Wales, Australia.
1992	The US FHWA states that: "Waste glass that is crushed and screened can be used as a portion of fine aggregate in asphalt paving mixes".
2020	Maxwell Lay, John Metcalf & Kieran Sharp prepare a book "Paving our ways: A History of the World's Roads and Pavements".

Bibliography

Abraham, H 1960, *Asphalts and allied substances*, vol. 1, historical review, Princeton: van Nostrand.

Agg, TR 1916 (5th edition 1940), *The construction of roads and pavements*, New York: McGraw-Hill.

Aitken, T 1900, *Road making and maintenance*, London: Charles Griffin.

Albert, W 1972, *The turnpike road system in England, 1663–1840*, Cambridge: Cambridge University Press.

Alcock, SE, Bodel, J & Talbert, RJA 2012, *Highways, byways and road systems in the pre-modern world*, Chichester: Wiley-Blackwell.

American Association of State Highway Officials 1961, *Interim guide for the design of rigid and flexible pavements*, Washington, DC: AASHO.

ANAS 2003, *l'Architettura delle Strade*, International Convention, Rome: ANAS SpA-formerly Azienda Nazionale Autonoma delle Strade (National Autonomous Roads Corporation).

Anderson, RMC 1932, *The roads of England*, London: Benn.

Anon 1795, *The life of John Metcalf, commonly called Blind Jack of Knaresborough*, York: Peck.

Anon 1856, 'Iron paving in London – replacement for granite setts', *Engineer*, vol. 1, 25 January, p. 48.

Anon 1876a, 'A decade of steam road-rolling in Paris', *Van Nostrand's Eclectic Engineering Magazine*, no. 6, pp. 298–302.

Anon 1876b, 'Asphalt roads in Paris', *Van Nostrand's Eclectic Engineering Magazine*, no. 6, p. 112.

Anon 1890, 'Slipperiness of pavements', *Engineering Record*, vol. 23, 25 January, pp. 113–28.

Anon 1891, 'Details of asphalt pavement work at Washington, DC', *Engineering Record*, vol. 25, 14 February, pp. 178–80.

Anon 1895, 'New pavements paragraph', *Scientific American*, vol. 73, no. 9, 31 August, p. 300.

Anon 1912, *Melbourne Punch*, 11 July, p. 6.

Anon 1913, 'A severe test of paving slabs', *figure 20.9*, pp. 303 & 305–7, June.

Anon 1915, 'The manufacture of granite paving blocks', *Engineering News*, vol. 73, 25 February, pp. 376–81.

Anon 1918–1924, *Public Roads*, vol. 1, no. 1, May; vol. 1, no. 2, June; vol. 5, no. 9, November 1924.

Anon 1926, 'An instrument for the measurement of relative road roughness', *Public Roads*, vol. 7, no. 7, pp. 144–51.

Anon 1930, 'New pavement has steel base', *ASCE Civil Engineering*, October, p. 53.

Anon 1933, 'A laboratory traffic test for low-cost road types', *Public Roads*, vol. 14, no. 1, p. 219.

Anon 1937, Herr Hitler's roads plan, *London Times*, 17 September.

Anon 1943, *Scientific American*, new pavements paragraph, vol. 169, no. 6, December.

Anon 1952, *Soil mechanics for road engineers*, London: HMSO.

Anon 1954, *Final report on Road Test One Maryland*, Special Report 4, Highway Research Board, Washington, DC.

Anon 1955 & 1962, *The WASHO Road Test, Part 1: design, construction and testing procedures*, Special Report 18, Washington DC, Highway Research Board, 1955 & *Part 2: Test data, analyses findings*, Special Report 22, Highway Research Board, Washington DC, 1962.

Anon 1980, 'History and progress: Georgia', *Better Roads*, November, pp. 16–8.

Anon 1989, 'The precursors', *PIARC Routes/Roads*, vol. 268, no. II.

Armstrong, EL (Editor) 1976, *History of public works in the United States 1776–1976*, Chicago, IL: American Public Works Association.

Attwooll, AW 1955, 'Rolled asphalt road surfacing materials', *Journal of the Institution of Highway Engineers*, vol. 3, no. 6, pp. 59–75.

Auff, AA 1982, *Quality control of dimensions in road construction*, SR 25, Vermont South: Australian Road Research Board.

Austroads 2004, *Pavement design: a guide to the structural design of road pavements*, Sydney, New South Wales: Austroads.

Austroads 2014, *Guide to pavement technology, part 4b: asphalt*, pub. no. AGPT04B-14, prepared by J Rebbechi & L Petho, Austroads, Sydney, New South Wales.

Austroads 2019a, *Guide to pavement technology, part 5: pavement evaluation and treatment design*, pub. no. AGPT05-19, prepared by G Jameson, Austroads, Sydney, New South Wales.

Austroads 2019b, *Guide to pavement technology, part 4d: stabilised materials*, pub. no. AGPT04D-19, prepared by R Andrews, Austroads, Sydney, New South Wales.

Autret, P, Baucheron de Boissoudy, A & Gramsammer, J-C 1987, 'The circular test track of the Laboratoire Central des Pants et Chaussées-Nantes, first results', *Conference internationale sur le dimensionnement des chaussées souplés*, Ann Arbor, Michigan, July.

Aveling-Barford Company 1965, *A hundred years of road rollers*, Lingfield, Surrey: Oakwood Press.

BC 1824, Improvement of the macadamized roads, Letter, *Mechanics Magazine*, vol. 3, July, p. 270.

Baker, IO 1903, *A treatise on roads and pavements*, New York: Wiley.

Bergier, N 1622, *Histoire des Grands Chemins de l'Empire Romain*, Paris: Morel. Reprint by Nabu Press, 2011.

Berthier, J 2013, 'de l'Essai AASHO au manège de fatigue', *Revue Générale des Routes et des Aérodromes*, vol. 914, September, pp. 22–27.

Bilkadi, Z 1996, *Babylon to Baku*, Windsor: Stanhope-Seta.

Bird, A 1969, *Roads and vehicles*, London: Arrow.

Bodey, H 1971, *Roads*, London: Batsford.

Borth, C 1969, *Mankind on the move*, Washington, DC: Automotive Safety Foundation.

Boulnois, HP 1895, *The construction of carriageways and footways*, London: Biggs.

Boulnois, HP 1919, *Modern roads*, London: Edward Arnold.

Boumphrey, G 1939, *British roads*, London: Thomas Nelson.

Bourne, D, 1763, *A treatise upon wheel-carriages; shewing their present defects*, London: Crowder.

Boussinesq, VJ 1885, *Application des potentiels à l'étude de équilibre et du movement des solides élastiques avec les notes étendues sur diverse pointes de physique, mathématique et d'analyse*, Paris: Gautier-Villais.

Boutet, D 1947, 'Deux grands ingénieurs des ponts et chaussées qui ont orienté leurs recherches dans le domaine routier', *Travaux*, vol. 152 (*La Route supplement*), June, pp. 9–14.

Bramley, RR 1805, *The road maker's guide*, Leeds: Wright.

Broome, DC 1963, 'The development of the modern asphalt road', *Surveyor and Municipal Engineer*, vol. 122, no. 3728, pp. 1437–40 & vol. 122, no. 3729, pp. 1472–75.

Buchanan, JA & Catudal, JA 1941, 'Standardizable equipment for evaluating road surface roughness', *Public Roads*, vol. 21, no. 12, pp. 227–44.

Burgoyne, JF 1843, *On rolling new-made roads*, Reproduced as Appendix I of Law & Clark 1907.

Burgoyne, JF c1860, *Remarks on the maintenance of macadamised roads*, Reproduced as Chapter 9 of Law 1850.

Burke, E (editor), *Annual Register 1775*, London: Longmans. Chronicle.

Burmister, DM 1943, 'The theory of stresses and displacements in layered systems and applications to the design of airport runways', *Proceedings Highway Research Board*, vol. 23, pp. 126–44.

Busch, T 2008, 'Connecting an empire; eighteenth century Russian roads, from Peter to Catherine', *Journal of Transport History*, vol. 29, no. 2, September, pp. 237–58.

Byrne, A 1917, *Modern road construction*, Chicago, IL: American Technical Society.

Carey, AE 1914, *The making of highroads*, London: Crosby Lockwood.

Carey, W & Irick, P 1962, The pavement serviceability performance concept /// 1973 Slope variance as a measure of roughness and /// The CHLOE profilometer, WN Carey, Jr, HC Huckins & RC Leather, Available at onlinepub.trb.org.

Carpenter, CA & Goode, JF 1936, 'Circular track tests on low-cost bituminous mixtures', *Public Roads*, vol. 17, no. 4, pp. 69–82 & vol. 19, no. 9, 1938 (cover photo).

Carpenter, CA & Willis, EA 1938, 'A study of sand-clay materials for basecourse construction', *Public Roads*, vol. 19, no. 9, pp. 173–90.

Catton, MD 1959, 'Early soil-cement research and development', *Proceedings ASCE*, vol. 85, no. HW1, pp. 1–16.

Chabrier, E 1876, *Proceedings Institution of Civil Engineers*, vol. XLIII, part 1, pp. 276–95.

Chamberlin, WP 1995, *Performance-related specifications for highway construction and rehabilitation*, National Cooperative Highway Research Program, Synthesis of Highway Practice 212.

Chartres, J 1981, 'L'homme et la route', *Journal Transport History*, vol. 2, no. 1, pp. 65–8.

Chevallier, R 1976, *Roman roads*, Berkeley: University of California Press.

Clark, J 1794, *General view of the agriculture of the county of Hereford with observations of the means of its improvement*, London: Colin McRae.

Clarke, D 1965, *The ingenious Mr Edgeworth*, London: Oldbourne.

Coane, JM, Coane, HG & Coane, JM Jr 1908 (1927, 4th edition with supplementary authors), *Australasian roads*, Melbourne: Robertson.

Codrington, T 1892, *The maintenance of macadamised roads*, 2nd edition, London: Spon.

Coffee, WF 1900, 'The "Orograph" an automatic profile recorder', *Scientific American*, vol. 82, no. 19, 12 May, pp. 293–4.

Coles, B & Coles, JM 1986, *Sweet track to Glastonbury: the Somerset levels in Prehistory*, London: Thames and Hudson.

Coles, JM 1984, 'Prehistoric roads and trackways in Britain', In *Loads and roads in Scotland and beyond*, Edited by A Fenton and G Stell, Edinburgh: John Donald.

Collingwood, F 1891, 'Street paving', *Engineering Record*, 7 March, pp. 223–4.

Collins, J & Hart, C 1936, *Principles of road engineering*, London: Edward Arnold.

Copeland, J 1968, *Roads and their traffic: 1750–1850*, Newton Abbot: David & Charles.

Corcoran, A 2006, *Road construction in Eighteenth century France*, HAL Centre pour la Communication Scientifique Directe, Lyon France.

Country Roads Board Victoria, 18th (1931) *Annual report*.

Crawford, C 1989, 'The rocky road of mix design', *Hot Mix Asphalt Technology*, Winter, pp. 10–6.

Crawford, DA 1938, 'Road construction by the heat-treatment of surface soils', *Proceedings VIIIth PIARC Congress*, The Hague, Section 1, Question 1, pp. 23–8.

Cresy, E 1847, *Encyclopaedia of civil engineering*, London: Longman.

Crompton, RE 1913, 'The mechanical engineering aspects of road construction', *Proceedings Institute of Mechanical Engineers*, 3/4, pp. 1253–1326 & 1928, Reminiscences, London: Constable.

Crosbie-Dawson, GJ 1876, 'Street pavements', *Journal of Liverpool Polytechnic Society*, vol. 270, pp. 16–75.

Darcy, HP 1850, 'Sur le pavage et le macadamisage des chaussées de Londres et de Paris', *Annales des Ponts et Chaussées*, 2e series, Tome, vol. 20, no. 233, pp. 1–58.

Davies, H 2002, *Roads in Roman Britain*, Gloucestershire: History Press.

de Camp, LS 1966, *The ancient engineers*, Bundy: Norwalk, Connecticut.

de Coulaine, Q 1850,' Sur l'emploir des substances bitumineuses dans la construction des chaussées, sur la nature, la composition, les propriêtés de ces substances, et leurs diverses applications', *Annales des Ponts et Chaussées*, Memoires et Documents, 2nd Cahier, Tome, vol. 19, pp. 240–308, Paris (also *Recherches* no. 230), In Malo 1866, pp. 190–5.

Deacon, GF 1879, 'Sheet carriageway pavements', *Proceedings Institution of Civil Engineers*, vol. 58, no. 4, pp. 1–30.

Defoe, D 1725, *A tour through the whole island of Great Britain*, vol 2. Appendix, (1962 reprint by Dent of London).

Delano, WH 1880, 'On the use of asphalt and mineral bitumen in engineering', *Proceedings Institution of Civil Engineers*, vol. 60, paper 1673, pp. 249–303.

Delano, WH 1893, *Twenty years practical experience of natural asphalt and mineral bitumen*, London: Spon.

Department of Main Roads New South Wales 1976, *The roadmakers: a history of main roads in NSW*, Sydney.

Descornet, G 1990, 'Reference road surfaces for vehicle testing', *PIARC Routes/ Roads*, vol. 272, no. 272.

Deveraux, R 1936, *John Loudon McAdam*, London: Oxford University Press.

d'Eyrinis, E 1721 (in Malo 1866).

Dickens, C 1856, 'A journey due north, Judy', *The London Serio-comic Journal*, 26 January 1870.

Dickinson, EJ 1984, *Bituminous Roads in Australia*, Vermont South: Australian Road Research Board.

Diodorus, S 54 BCE, *Library of history*.

Distin, P 1992, 'Surface dressing – how it all came about', *Highways and Transportation*, vol. 39, no. 8, September, pp. 5–8.

DKC 1878, 'Apparatus for measuring the comparative strength of broken stones', *Minutes of Proceedings of Institution of Civil Engineers*, vol. 57, no. 3, p. 31.

Dohmen, LJM & Molenaar, AAA 1992, Full scale pavement testing in the Netherlands, *Proceedings on 7th International Conference on Asphalt pavements*, Nottingham UK, vol 2, pp. 64–82.

Doppler, HW 1980, 'Alte Strassen über den Bözberg AG', *Archäologie der Schweiz*, vol. 3, no. 3, pp. 1–4.

du Plessis, L, Coetzee, NF, Burmas, N, Harvey, JT & Monismith, CL 2008, 'The Heavy Vehicle Simulator in accelerated pavement testing – a historical overview and new developments', *Proceedings Accelerated Pavement Testing (APT) 3rd International Conference*, Madrid, Spain, 1–3 October, pp. 17–33.

Dunham, AL 1955, *The industrial revolution in France, 1815–1848*, New York: Exposition Press.

Earle, JB 1971, *A century of roadmaking materials*, Oxford: Blackwell.

Earle, JB 1974, *Blacktop*, Oxford: Blackwell.

Edgeworth, RL 1817, *Essay on the construction of roads and carriages, etc.* 2nd edition, London: Hunter.

Ekse, M & LaCross, LM 1957, 'Model analysis of flexible pavement and subgrade stresses', *Proceedings Association of Asphalt Paving Technologists*, vol. 2, pp. 312–20.

Ellice-Clark, EB 1880, 'Asphalte and its application to street paving', *Engineering News*, no. 1 May, pp. 154–7.

Ellis, PB 2003, *A brief history of the Celts*, London: Robinson.

Emmons, WJ 1934, *Public Roads*, vol. 14, no. 11, January, pp. 197–218.

ERPUG 2015, History of road profile measurements and current standards in road surface characteristics, Leif Sjögren, VTI Federal Highway Administration, 1976, *America's Highways 1776–1976*, Washington: US Dept Transportation<www.erpug.org/media/files/forelasningar_2015/1_ERPUG_2015_Leif_VTI.pdf, PDF file, Budapest, Hungary 2015>.

Foote, C 1994, Letter, *Highways and Transportation*, vol. 41, no. 4, January.

Forbes, FJ 1953, 'Roads of the past', *Chemistry and Industry*, vol. 4, 24 January, pp. 70–4.

Forbes, FJ 1958a, Roads to c1900, In *A history of technology*, edited by C Singer, et al., vol. 4, pp. 520–47, Oxford: Oxford University Press.

Forbes, R 1958b, Petroleum, In *A history of technology*, edited by C Singer, et al., vol. 5, pp. 102–23, Oxford: Oxford University Press.

Forbes, RJ 1934, *Notes on the history of ancient roads and their construction*, Amsterdam: NV Noord-Hollandsche, Archaeologisch-historische Bijdragen, Deel III, University of Amsterdam.

Forbes, RJ 1936, *Bitumen and petroleum in antiquity*, Leiden: Brill.

Fortidsnindeforvaltning undated, The "Broskov" roads, ancient and medieval roads excavated 1953–61, Copenhagen.

Fortune, W 1893, 'Street paving in America', *Century Magazine*, vol. 46, no. 24, May–October.

Franklin Institute, Committee on Science and the Arts 1843, 'Report on the best modes of paving highways', *Journal of Franklin Institute*, vol. 6, series 3(3), September, pp. 145–68; October, pp. 217–33, or vol. 36, no. 3.

French, D 1981, *Roman roads and milestones of Asia Minor, Fasc. 1: The Pilgrims Road*, British Institute of Archaeology at Ankara, Monograph no. 3, BAR International Series 105.

Gautier, H 1693, *Traité de la construction des chemins*, Paris: Cailleau.

Gerke, R 1987, *Subsurface drainage of road structures*, SR35, Vermont South: Australian Road Research Board.

Giles, GC & Sabey, BE 1959, 'A note on the problem of seasonal variation in skid resistance', *Proceedings 1st International Skid Prevention Conference*, Part 2, pp. 563–8.

Gillespie, HM 1847, *A manual of the principles and practice of road-making*, New York: Barnes.

Gillespie, HM 1992, *A century of progress: the history of hot mix asphalt*, Lanham, MD: National Asphalt Paving Association.

Gillespie, TD, Paterson, WDO & Sayers, MW 1986, *Guidelines for conducting and calibrating road roughness measurements*, Technical Paper 46, World Bank, Washington, DC.

Gillmore, QA 1876, *A practical treatise on roads, streets and pavements*, New York: Van Nostrand.

Green, JW & Ridley, CN 1927, *The science of roadmaking*, London: Crosby Lockwood.

Greene, FV 1885, 'Pavements here and abroad', *Engineering News and American Contract Journal*, vol. XV, 19 September, p. 186.

Gregory, JW 1931, *The story of the road*, London: Alexander Maclehose.

Griffin, E 1887, 'The streets of Washington, DC in 1886–87', *Engineering News*, vol. 22, October, pp. 297–8.

Gross, H 1901, 'The evolution of street paving in Chicago', *Municipal Journal & Engineer*, vol. 11, November, pp. 206–7.

Grübler, A 1990, *The rise and fall of infrastructures*, Heidelberg: Physica-Verlag.

Guldi, J 2012, *Roads to power*, Cambridge, MA: Harvard University Press.

Haas, R, Hudson, WR & Falls, C 2005, *Pavement asset management*, New York: Wiley.

Hadfield, WJ 1934, *Highways & their maintenance*, London: The Contractors' Record.

Haldane, A 1962, *New ways through the Glens*, Argyll, Scotland: House of Lochar.

Halladay, M 1998, *Public Roads*, vol. 1, no. 5, March-April <www.fhwa.dot.gov/publications/publicroads>.

Halstead, WJ & Welborn, JY 1974, 'History of the development of asphalt testing apparatus and asphalt specifications', *Proceedings Association of Asphalt Paving Technologists*, vol. 43A, pp. 89–120.

Hanson, FM 1935, 'Bituminous surface treatment of rural highways', *Proceedings NZ Society of Civil Engineers*, vol. 21, pp. 81–179.

Harman, T, Bukowski, JR, Moutier, F, Huber, G & McGennis, R 2002, 'History & future challenges of gyratory compaction', *Transportation Research Record*, vol. 1789, pp. 200–7.

Harrell, JA. & Bown, TM 1995, 'An Old Kingdom basalt quarry at Widan el-Faras and the Quarry Road to Lake Moeris', *Journal of American Research Center in Egypt*, no. 32, 71–91.

Hartman, C 1927, *The story of the roads*, London: G Routledge.

Haywood (not Heywood), W 1871, *Report to the streets committee of the Hon. the Commissioner of Sewers of the City of London, upon granite and asphalte pavements*, London: Judd.

Herodotus, The Histories, Book 2, In RB Strassler's 'The Landmark Herodotus', New York: Pantheon, 2007.

Hey, D 1980, *Packmen, carriers and packhorse roads*, Leicester: Leicester University Press.

Highway Research Board 1961, *The AASHO Road Test: Report 7 – Summary Report*, Special Report 61G. Washington, DC.

Hindley, G 1971, *A history of roads*, London: Davies.

Hines, EN 1913, *Concrete roads of Wayne County*, Portland Cement Association.

Hoffbeck, SR 1991, 'The pavement problem', *Invention & Technology*, vol. 7, no. 2, pp. 16–7.

Hogentogler, CA 1923, 'Apparatus used in highway research projects in the United States', *Bulletin*, vol. 6, no. 4, p. 35, Washington: US National Research Council.

Hoiberg AJ 1965, *Bituminous materials*, Vol II, asphalts, New York: Krieger.

Holley, IB Jr 2003, 'Blacktop: how asphalt paving came to the urban United States', *Technology and Culture*, vol. 44, no. 4, October, pp. 703–733.

Hosking, R 1992, *Road aggregates and skidding*, TRL State-of-the-art review 4, London: HMSO.

Hubbard, P 1910, *Dust preventives and road binders*, New York: John Wiley and Sons.

Hudson, RW & Hain, RC 1961, Calibration and Use of BPR Roughometer at the AASHO Road Test, Highway Research Board Special Report 66.

Huft, D 2010, *A trip through time: it's been a bumpy road*, Roanoke, VA: South Dakota Dept of Transportation Pavement Evaluation.

Hughes, SA, Adam, W & China, F 1938, *Tar roads*, London: Edward Arnold.

Hugo, F & Epps-Martin, AL, 2004, *Significant findings from full-scale accelerated pavement testing*, NCHRP Synthesis of Highway Practice 325, Washington DC: Transportation Research Board.

Hugo, V 1862, *Les Misérables*, Penguin Classics (Denny translation).

Hveem, F. 1955, 'Pavement deflections and fatigue failures', *Highway Research Board*, Bulletin 114, pp. 43–87.

Hveem, FN 1960, 'Devices for recording and evaluating pavement roughness', *Highway Research Board*, Bulletin, p. 264.

Hveem, FN 1971, 'Asphalt pavements from the ancient east to the modern west', *Highway Research News*, no. 42, Winter, pp. 21–39.

Hveem, FN & Carmany, RM 1948, 'The Factors underlying the rational design of pavements', *Proceedings Highway Research Board*, December, pp. 7–10.

Hyslop, J 1984, *The Inka road system*, Orlando: Academic Press.

Ingles, OG & Metcalf, JB 1972, *Soil stabilisation*, Sydney: Butterworths.

Ioannides, AM & Khazanovich, L 1993, 'Load equivalency concepts: a mechanistic reappraisal', *Transportation Research Record*, vol. 1388, pp. 42–51, TRB, Washington

Ironbridge Gorge Museum Trust, 1979 edition, *The Tar Tunnel*, Museum Guide 4.04.

Jackman, WT 1916, *The development of transportation in modern England*, Cambridge, MA: Cambridge University Press.

James, JG 1964a, 'History of bituminous surfacings', Letter, *Surveyor and Municipal Engineer*, vol. 123, no. 3736, pp. 25–6.

James, JG 1964b, '50 years of white lines', *Roads and Road Construction*, vol. 42, December, pp. 409–14.

Jameson, GW 1996. *Origin of the Austroads design procedures for granular pavements*, ARR 292, Vermont South: Australian Road Research Board.

Japan Road Association 2010, *Road pavements in Japan: technical standard and latest technology*, prepared by K Endo, S Horiuchi, M Iwama, O Kamada, & M Shimazaki.

Japan Road Association 2012, 'From ancient times to the Edo period: birth and development of the first road network in Japan', *Routes-Roads*, vol. 356, 4th quarter, pp. 81–5.

Jordan, RP 1985, 'Viking trail east', *National Geographic*, vol. 167, no. 3, pp. 278–317.

Judson, WP 1908, *Road preservation and dust prevention*, New York: The Engineering News Publishing Co.

Jusserand, J 1901, *English wayfaring life in the Middle Ages (XIVth Century)*, London: Unwin.

Kane, V undated, *Design for the next generation of pavements: flexible pavement design*, IRF Webinar 170426, undated.

Kellett, JR 1969, *The impact of railways on Victorian cities*, London: Routledge & Kegan Paul

Kennerell, EJ 1958, 'Roads from the beginning', *Journal of Institution of Highway Engineers*, vol. 5, no. 3, pp. 176–205.

Kerzého, J-P 2013, 'Set-up and use of the IFSTTAR's fatigue test track', *RGRA European Road Review*, Fall, pp. 22–31 (includes paper on pavement design and up-grading).

Kier, WG 1974, Bump-integrator measurements in routine assessment of highway maintenance needs, TRRL Supplementary Report 26UC.

Kirkaldy, AW & Evans, AD 1920, *The history and economics of transport*, 2nd edition, London: Sir Isaac Pitman.

Kirwan, RJ 1911, Second Irish Road Congress.

Koldewey, R 1914, *The excavations at Babylon*, London: Macmillan.

Krchma, LC & Gagle, DW 1974, 'A USA history of asphalt refined from crude oil and its distribution', *Proceedings Association of Asphalt Paving Technologists*, 43A (50th Anniversary Historical Review), pp. 25–88.

Ksaibati, K, Whelan, M & Burczyk, J 1994, *Selection of subgrade modulus for pavement overlay design*, Laramie: University of Wyoming.

Law, H 1850, *Rudiments of the art of constructing and repairing common roads*, London: John Weale.

Law, H & Clark, DK 1901, *The construction of roads and streets*, London: Crosby Lockwood and Sons 6th edition. & (7th edition 1907).

Law, WM 1962, 'The development and use of coated macadam', *Proceedings Institution of Works and Highways Superintendents, Annual Conference*, Brochures, Hastings.

Lay, MG 1984, *History of Australian roads*, SR 29, Vermont South: Australian Road Research Board.

Lay, MG 1993, 'Modelling pavement behavior', *Road and Transport Research*, vol. 2, no. 2, June, pp. 16–27.

Lay, MG 1996, 'The history of compaction, Part 2, the sheepsfoot roller', *PIARC Routes/Roads*, vol. 290, no. II, pp. 44–50.

Lay, MG 1997, 'The Metcalf inheritance', *PIARC Routes/Roads*, vol. 293, no. 1, pp. 65–71.

Lay, MG 1999, 'One of our models is missing', *Australian Road Research*, vol. 8, no. 1, March, pp. 96–8.

Lay, MG 2001, 'Private sector funding and management of road infrastructure', *Proceedings 20th Australian Road Research Board Conference (Melbourne)*, invited paper, pp. 45–56.

Lay, MG 2009, *Handbook of road technology*, 4th edition, London: Spon.

Lay, MG 2010, *Strange ways*, Crows Nest, New South Wales: EA Books.

Lay, MG 2018, *The harnessing of power*, Newcastle upon Tyne: Cambridge Scholars Publishing.

Leahy, E 1844, *A practical treatise on the making and repairing of roads*, London: Weale.

Leighton, AC 1972, *Transport and communications in early medieval Europe: AD 500–1100*, Newton Abbot: David & Charles.

Leupold, J 1725, *Theatrum machinarum*, Leipzig: Buntel.

Lochhead, W 1878. Letter to the Editor, *Engineer*, no. 46, 15 November, p. 358.

MacAlister, RAS 1932, An ancient road in the Bog of Allen', *Journal of the Royal Society of Antiquaries of Ireland*, XLII, Part 2.

Macaulay, TB 1849, *The history of England from the accession of James the Second*, vol. 1, part 3, London: Longmans Green and Co.

Macneill, J 1833, 'Road indicator or instrument for ascertaining the comparative merit of roads and the state of repair in which they are kept', London: Roake and Varty. *The instrument is described in the Seventh report of the Commissioners appointed under the acts of 4 geo. iv. c. 74 1830*, Holyhead, pp. 20–23.

Mahoney, J P & Terrel, RL 1982, 'Laboratory and field fatigue characterisation for sulphur extended asphalt paving mixtures', *Proceedings Fifth International Conference on the Structural Design of Asphalt Pavements*, University of Michigan, Ann Arbor, pp. 831–43.

Mayhew, H 1850, 'London labour and London poor', In P. Quennell, *Mayhew's London*, London: Spring Books.

Malcolm, L 1933, 'Early history of the streets and paving of London', *Transactions of Newcomen Society*, vol. 14, pp. 83–4, p. 85.

Malo, L 1866, *Guide practique pour la fabrication et l'application l'asphalte et des bitumes*, E Lacroix, editor, Paris: Libraire Scientifique, Industrielle et Agricole (their Series D, No 18).

Malo, L 1886, *On asphalte roadways*, London: Spon.

Manton, BG 1956, 'John Loudon McAdam, born September 21st, 1756', *Highways and Bridges and Engineering Works*, vol. 24, no. 1157, pp. 6–10.

Markwick, AH & Shergold FA 1945, 'The aggregate crushing test for evaluating the mechanical strength of course aggregates', *Journal of the Institution of Civil Engineers*, vol. 24, no. 6, pp. 125–33.

Martin, D 2005, 'A font of ideas, and his stuck fast', Obit. *Melbourne Age*, 2 November.

Martin, TC 1996, *A review of existing pavement performance relationships*, ARR 282, Vermont South: Australian Road Research Board.

Mather, W 1696, *On the repairing and mending of highways*, England: Pamphlet.

McAdam, JL 1816, *Remarks (or observations) on the present system of roadmaking*, 1st edition, London: Longman, Rees, Orme, Brown and Green.

McAdam, JL 1819, *A practical essay on the scientific repair and preservation of public roads*, London: the author.

McAdam, JL 1824, *Remarks on the present system of road making; with observations deduced from practice and experience, etc*, 8th edition, London: Longman, Hurst, Rees, etc.

McAdam, JL 1825, *Observations on the management of trusts and the care of turnpike roads*, London: Longmans.

McCloskey, JF 1949, 'History of military road construction', *Military Engineer*, vol. XLI, no. 283, pp. 353–56.

McNichol, D 2005, *Paving the way: asphalt in America*, Lanham, MD: National Asphalt Paving Association.

McShane, C 1979, Transforming the use of urban space: a look at the revolution in street pavements, 1880–1924', *Journal of Urban History*, vol. 5, no. 3, pp. 297–307.

McShane, C 1994, *Down the asphalt path: the automobile and the American city*, New York: Columbia University Press.

Merdinger, CJ 1952. 'Roads – through the ages', *Journal Society of American Military Engineer*, vol. 44, no. 300, pp. 268–73 & vol. 44, no. 301, pp. 340–4.

Metcalf, JB 1996, *Application of full-scale accelerated pavement testing*, Synthesis of Highway Practice 235, Washington, DC: Transportation Research Board.

Metcalf, JB 2014, 'A history of full-scale accelerated pavement testing facilities to 1962', *Road and Transport Research*, vol. 23, no. 4, pp. 25–40.

Meyn, L 1873, 'The asphalts', *Van Nostrand's Eclectic Engineering Magazine*, pp. 74–82 & 123–6.

Moaligou, C 1982, 'Le pavé de Paris', *Revue Générale des Routes at des Aérodromes*, vol. 583, pp. 39–43.

Moch, J 1947, *Discours Travaux*, no. 153, pp. 306–7, July.

Monismith, CL & Brown, SF 1999, 'Developments in the structural design and rehabilitation of asphalt pavements over three quarters of a century', *Asphalt Paving Technology*, vol. 68A, March, pp. 128–251.

Monismith, CL & Witczak, MW 1983, 'Moderators' report. Papers in session 1. Pavement design', *Proceedings 5th International Conference on the Structural Design of Asphalt Pavements*, Delft 1982, University of Michigan, vol. 2, p. 2.

Moore, R 1876, 'Street pavements: an exhaustive review', *Engineering News*, vol. 6, May, p. 146, 20 May, p. 163, 170; 3 June, p. 179 (from the St Louis Republican).

Morin, A 1842, *Expériences sur le tirage des voitures et sur les effets destructeurs qu'elles exercent sur les routes*, Paris: Ministre des Travaux Publics.

Morris, PO. 1972, *The performance of pavement materials on the grey-brown soils of western Queensland*, AIR 013-1, Vermont South: Australian Road Research Board.

Mountain, AC 1880, *Report on paved roadways*, City of Sydney.

Mountain, AC 1885, *Reply to the Wood-paving Board report*, Sydney: City Surveyor's Office, March.

Mountain, AC 1887, 'Paved carriageways in Sydney, New South Wales', *Proceedings Institute of Civil Engineers*, vol. XCIII, part iii.

Mountain, AC 1894, *The evolution of the modern road*, Melbourne: Victorian Institute of Engineers.

Mountain, AC 1903, *Wood paving and road making in Australia*, second issue, Melbourne: Modern Printing.

Mountain, AC 1913, 'Wood paving report', *PIARC London Congress*, Sub-section B.4, report 28b.

Mullins, HB 1859, 'An historical sketch of engineering in Ireland', *Transactions Institution of Civil Engineers of Ireland*, vol. VI.

National Association of Australian State Road Authorities 1987, *Pavement design*, Sydney, New South Wales: NAASRA.

Needham, NJ 1971, 'Science and civilisation in China', *Civil engineering and Nautics*, vol. 4, no. 3, Cambridge: Cambridge University Press.

Newcomen, T 1951, 'Pre-Roman roads in Britain', *Contractors Record and Municipal Engineering*, 21 February, pp. 11–14, 28.

Newlon, H Jr, Pawlett, NM, et al. 1985, *Backsights*, Richmond, VA: Virginia Dept of Highways and Transportation.

North, EP 1879, 'The construction and maintenance of roads', *Transactions ASCE*, vol. 8, pp. 95–147, May.

Nylan, M 2012, 'The power of highway networks during China's classical era (323 BCE – 316 CE)', In *Alcock* 2012.

Oraganisation for Economic Cooperation and Development 1985, *Full-scale pavement tests*, Paris: OECD.

Office of Public Roads 1913, Bulletin 47, March, Washington, DC: US Federal Government.

O'Keeffe, PJ 1973, 'The development of Ireland's road network', *Proceedings Institution of Engineers of Ireland*, Civil Division papers.

Parnell, H 1838, *A treatise on roads*, London: Longman, Orme, Brown, Green and Longmans.

Parsons, WB 1939, *Engineers and engineering in the Renaissance*, Cambridge, MA: MIT Press.

Paterson, J 1819, *A practical treatise on the making and upholding of public roads*, Dundee: J. Watt.

Pawson, E 1977, *Transport and economy: the turnpike roads of eighteenth century Britain*, London: Academic Press.

Peckham, SF 1909, *Solid bitumens: their physical and chemical properties*, New York: Myron Clark.

Permanent International Association of Road Congresses 1923, 4th Congress Seville, 1st session, 'Surfacing of roads with concrete'.

Phillips, R 1737, A dissertation concerning the present state of the high roads of England, Presented to the Royal Society. Printed and sold by L. Gilliver and J. Clarke at Homer's Head in Fleetstreet and at their shop in Westminster Hall; J Stephens at the Hand and Star between the Temple-Gates and Fleetstreet; and J. Roberts in Warwick Lane. (1903 University of London copy)

Pierre F & Cohrs, H-H 1998, *The history of road building equipment*, Wadhurst: KLH International.

Porter, OJ 1938, 'The preparation of subgrades', *Proceedings Highway Research Board*, vol. 18, no. 2, pp. 324–31.

Porter, OJ 1949. 'Development of the original method for highway design', *Proceedings ASCE*, vol. 75, pp. 11–7.

Portland Cement Association (US) 1984, Thickness design for concrete highway and street pavements, EB 109P.

Powell, WD, Potter, JF, Mayhew, HC & Nunn, ME 1984, *The structural design of bituminous roads*, TRRL Laboratory Report 1132.

Prem, H 1989, *NAASRA roughness meter calibration via the road-profile based International Roughness Index (IRI)*, ARR 164, Melbourne, Victoria: Australian Road Research Board.

Proctor, T 1607, *A worthy worke profitable to this whole Kingdome concerning the mending of all high-waies*, London: Allde (1977 facsimile edition).

Ransom, PJ 1984, *The archaeology of the transport revolution*, Tadworth: World's Work.

Ravensteijn, W & Kop, J (editors) 2008, *For profit and prosperity*, Netherlands: KITLV Press, ISBN: 9789059942219. Now published by Brill.

Ray, GK 1964, 'History and development of concrete pavement design', *Journal of Highway Division, Proceedings ASCE*, vol. 90, no. HW1, pp. 79–101

Reader, WJ 1980. *McAdam*, London: Heineman.

Reilly, PC 1918, 'Destruction of wood block pavement due to use of tar in the creosote oil', *Municipal Engineering*, vol. LIV, no. 4, May, pp. 139–40.

Report 1886, 'The report of the executive board of Rochester, New York', *Engineering News and American Contract Journal*, vol. 16, 20 November, pp. 328–30.

Reverdy, G 1980, 'Les chemins de Paris en Languedoc. Part 5 of Histoire des grandes liaisons françaises', *Revue Générale des Routes et des Aérodromes*, no. 583.

Revue Générale des Routes et des Aérodromes 1976, *Cinquante ans au service de la Route*, Special edition

Richardson, C 1905, *The modern asphalt pavement*, New York: John Wiley (2nd edition, 1908).

Rickman, J (editor) 1838, *Life of Thomas Telford: civil engineer: written by himself*, London: Hansard.

Ritchie, B 1999, *The story of tarmac*, London: James & James.

Road Research Laboratory 1962, *Bituminous materials in road construction*, London: HMSO.

Robertson, JN 1958, 'The pavement of presidents', *Asphalt Institute Quarterly*, vol. 10, no. 2, pp. 4–7.

Rolt, LC 1958, *Thomas Telford*, London: Longmans.

Rose, AC 1952, *Public roads of the past: 35 BC to 1800 AD*, Washington, DC: AASHO.

Rose, AC 1953, *Historic American highways*, Washington, DC: AASHO.

Rose, G & Bennett, D 1994, 'Benefits from research investment: case of Australian Accelerated Loading Facility pavement research program', *Transportation Research Record*, vol. 1455, pp. 82–90, Washington, DC: National Research Council.

Sachs, W 1992 (1984 in Germany), *For love of the automobile*, Berkeley: University of California Press.

Sauterey, R 1997, 'A brief historical chronicle of PIARC', *Routes/Roads*, vol. 296, no. IV, pp. 76–9.

Sayers, MW 1995, 'On the calculation of International Roughness Index from longitudinal road profile', *Transportation Research Record*, vol. 1501, pp. 1–12.

Sayers, MW, Gillespie, TD & Queiroz, CAV 1986, The International Road Roughness Experiment (IRRE): establishing correlation and a calibration standard for measurements, World Bank Technical Paper 45. Washington, DC.

Sayers MW & Karamihas SM 1998, *The little book of profiling*, Ann Arbor, MI: Michigan University.

Scala, AJ 1970, 'The use of the PCA Roadmeter for measuring road roughness', *Proceedings 5th Australian Road Research Board Conference*, vol. 5, no. 4, pp. 348–67.

Schreiber, H 1961, *The history of roads*, London: Barrie and Rockliff.

Scott-Giles, CW 1946, *The road goes on*, London: Epworth.

Seely, BE 1984, 'The scientific mystique in engineering: highway research at the Bureau of Public Roads, 1918–1940', *Technology and Culture*, vol. 25, no. 4, October, pp. 798–831.

Shackel, B & Arora, N 1978, 'The application of a full-scale road simulator to the study of highway pavements', *Australian Road Research*, vol. 8, no. 2, pp. 17–31.

Sharp, KG 2004, 'Full scale accelerated pavement testing: a southern hemisphere and Asian perspective', *Proceedings 2nd International Conference on Accelerated Pavement Testing*, Minneapolis, Minnesota, USA.

Sharp, KG & Armstrong, PJ 1985, *Interlocking concrete block pavements*, SR 31, Vermont South: Australian Road Research Board.

Sharp, KG, Brindle, RE, Lay, MG, Metcalf, JB, Johnston, IR & Jones, D 2011, *ARRB the first fifty years*, Melbourne, Victoria: Australian Road Research Board.

Sharp, KG, Patrick, S, Thananjeyan, A & Simpson, C 2012, *Field validation of warm mix asphalt pavements*, publication no. AP-T214/12, November, Sydney: Austroads.

Shell International Petroleum Ltd 1978, *Shell pavement design manual*, London: Shell.

Smiles, S 1874, *Lives of the engineers, history of roads, Metcalfe*, Telford, London: John Murray.

Southey, R 1829, *Journal of a tour in Scotland in 1819*, London: John Murray.

Sparks, GH & Davis, EH 1970, 'First road base experiment with a laboratory test track', *Proceedings 5th Australian Road Research Board Conference*, vol. 5, no. 4, pp. 284–301.

Sparks, GH, Taylor, H & Lee, IK 1968, 'Development and instrumentation of the model road test track at Sydney University', *Proceedings 4th Australian Road Research Board Conference*, vol. 4, no. 2, pp. 1341–55.

Spencer, H 1965. History of asphalt paving. In *Hoiberg* 1965.

Spielmann, PE & Hughes, AC 1936, *Asphalt roads*, London: Arnold.

St George, PW 1879, 'On the street and footwalk pavements of Montreal, Canada, from the year 1842 to 1878', *Proceedings Institution of Civil Engineers*, vol. 58, no. 4, pp. 287–309.

Stevenson, R 1824, *Roads and highways: entry in the Edinburgh Encyclopaedia*, Edinburgh: Balfour.

Talbot, AN & Richart, FE 1923, The strength of concrete and its relation to the cement, aggregate, and water, Illinois University Bulletin, no. 137.

Taylor, C 1979, *Roads and tracks of Britain*, London: Dent.

Teller, L & Sutherland, EC 1935, *Public Roads*, vol. 16, no. 8, p. 145 & vol. 16, no. 10, p. 201.

Terrel, RL & Krukar, M 1969, 'Evaluation of test tracking pavements', *Proceedings Association of Asphalt Paving Technologists*, vol. 38, pp. 272–96.

von Terzaghi, K 1960, *From theory to practice in soil mechanics*, New York: Wiley.

The Holy Bible, Revised Standard version, 1973, New York: Collins.

Thomson, JT 1928, 'Effect of pavement type on impact reaction', *Public Roads*, vol. 9, no. 6, pp. 113–22.

Tillson, GW 1897, 'Asphalt and asphalt pavements', *Transactions ASCE*, vol. 38, December, pp. 215–35.

Tillson, GW 1901, *Street pavements and paving materials*, New York: Wiley.

Toglietta, G 1585, 'Discorso del mattonato e selicato di Roma', *Ms Cart. del Sec.*, vol. XVI, Biblioteca Nazionale Vittorio Emanuele, Rome (Reprinted in 1878 by Real Societa di Storia Patriae, I:371).

Towler, J & Dawson, J 2008, 'History of chip sealing in New Zealand – Hanson to P/17', *Proceedings Sprayed Sealing Conference*, Vermont South: Australian Road Research Board.

Transport and Road Research Laboratory 1960, *A guide to the structural design of pavements for new roads*, Road Note RN 29. The last Edition was published in 1970.

Transportation Research Board 2007, 'Pavement lessons learned from the AASHO Road Test and performance of the Interstate Highway System', *TRB Circular*, E-C118, July.

Trésaguet, P-M 1831, 'Memoire sur l'entretier des chemins de la generalité des Limoges', *Annales des ponts et chausées*, 1st series, vol. 2, pp. 243–58.

Tunnicliff, DG, Beaty, RW & Holt, EH 1974, 'A history of plants, equipment and methods in bituminous paving', *Proceedings Association Asphalt Paving Technologists*, vol. 43A (50th Anniv Hist Revue), pp. 159–296.

Ullidtz P 1982, 'Predicting pavement response and performance from full scale testing', *Proceedings International on Full Scale Pavement Tests*, Institut Fur Strassen, Eisenbahn und Felsbau an der Eidigenossischen Technischen Hochschule Zurich.

Vahrenkamp, R 2010, *The German autobahn 1920–1945*, Lohmar: EUL Verlag.

Vuong, BT 2008, 'Validation of a three-dimensional nonlinear finite element model for predicting response and performance of granular pavements with thin bituminous surfaces under heavy vehicles', *Road and Transport Research*, vol. 17, no. 1, pp. 3–31.

Warren, F 1901, 'The development of bituminous pavements', *Engineering Record*, vol. 44, no. 9, 31 August, pp. 198–9.

Watanatada, T, Harral, C, Paterson, W, Bhandari, A & Tsunokawa, K. 1987. *The highwav design and maintenance standards model*, 2 vols, Baltimore, MD: The Johns Hopkins University Press.

Webb, S & Webb, B 1913, *English local government: the story of the King's Highway*, London: Longmans.

Weingroff, RF 2008, 'A journey to better highways: 100 years of public roads', *Public Roads*, vol. 82, no. 2, Summer www.fhwa.dot.gov/publications/publicroads/18summer/06.cfm.

Westergaard, HM 1926, 'Stresses in concrete pavements computed by theoretical analysis', *Public Roads*, vol. 7, no. 2, pp. 25–35.

White, G 1789, *The natural history and antiquities of Selborne*, London: White.

Williams, DH 1998, 'Cistercian roads and route-ways', *Tarmac Papers*, vol. II, pp. 231–46.

Wooltorton, FL 1954, *The scientific basis of road design*, London: Edward Arnold.

Young, A 1770, *A six months' tour through the North of England*, 4 volumes, London: Strahan.

Young, A 1809, *Agricultural survey of Oxfordshire*, London: Phillips.

End Notes

1 Cartoon with kind permission of the Leunig Studio.
2 Scott-Giles 1946, p92.
3 See Lay 2009 for information on those matters.
4 Lay 1992, p9.
5 Deacon 1879, p1.
6 Franklin Institute 1843.
7 Mountain 1880, p40.
8 Hveem & Carmany 1948, p102.

CHAPTER 1

9 From a poem in Silvae by Stratius, translation in Chevallier 1976, p83.
10 Lay 2009, Section 11.1.1.
11 e.g. Lay 2009.
12 Lay 2009, p3.
13 Lay 2009, Chapter 26.
14 Haldane 1962, Chapters 1 and 2.
15 Gerke 1987.
16 Lay 2009, Section 9.5.
17 Lay 2009, Chapter 9.
18 Jusserand 1901, p85.
19 Parsons 1939, p302.
20 Lay 2010, p51.
21 Scott-Giles 1946, p50.

CHAPTER 2

22 Harrell & Bown 1995, pp71–91.
23 Forbes 1934, p60, discusses remnants of a similar road a further 50 km south.
24 Herodotus, p174.
25 Hindley 1971, p20.

26 Herodotus, p180.
27 Many sources are mapped in Abraham 1960. Further data on the history of bitumen is outlined in Lay 1992, pp242–3.
28 Translations are from the 1973 edition of The Holy Bible, 1973.
29 The Holy Bible 1973.
30 Bilkadi 1996, p45.
31 Lay 1992, p50; Abraham 1960, p23.
32 Aitken 1900, p4.
33 de Camp 1966, p71.
34 Abraham 1960, p24.
35 Koldewey 1914, p54.
36 Forbes 1936, p69.
37 Abraham 1960, Figure 25.
38 Diodorus 54 BCE, 19.2; see also Hindley 1971, p21.
39 Schreiber 1961, p103.
40 via Forbes 1934, pp50–1.
41 Forbes 1934, p55.
42 Photo by Jenn Funk Weber at jenfunkweber.com.
43 Chevallier 1976, p89.
44 Forbes 1934, p102.
45 Forbes 1934, p48. Schreiber 1961, p20, 99–100, 112, 116.
46 Nylan, M. 2012. p37.
47 Doppler 1980, 3:1, and photo with permission of the Aargau Kantonsarchaeologie.
48 Doppler 1980, p26.
49 Chevallier 1976, pp12–5.
50 Schreiber 1961, p116.
51 Maurus Sevius, Libros 15, Originum. Sourced in Gregory 1931, p60.

CHAPTER 3

52 Hey 1980, p64.
53 Byrne 1917, p77.
54 Aitken 1900, p15.
55 Byrne 1917, p80.
56 Federal Highway Administration 1976, p72.
57 Photo courtesy of Victorian Department of Transport.
58 VolvoCE website.
59 Byrne 1917, p84.
60 Byrne 1917, pp84–5.
61 From NAASRA 1987, Figure 4.1. Published with permission of Austroads.
62 An Act for the continuance of the statute made 2 & 3 P. & M. for the amendment of highways. 5 Elizabeth, Cap XIII:555 (VI). 1562.
63 Proctor, T. 1607.
64 Macaulay 1849, Chapter 5.
65 Lay 2018, Chapter 4c.

66 Deveraux 1936, p9.
67 Phillips 1737, p9.
68 Kirkaldy & Evans 1920.
69 Young 1770, p581.
70 Lay 1992, p66; Lay 2010, p59.
71 Newlon & Pawlett 1985, p48.
72 Schreiber 1961, p49.
73 White, G 1789, letter V.
74 Smiles, S 1874.
75 Hey, D 1980.
76 Haldane 1962, p6.
77 Phillips, R. 1737.
78 Addison 1980, p115.
79 Aitken 1900, p10.
80 Lay 1992, p67.
81 Third Report form the (British) Parliamentary Report on Turnpikes and Highways. 1809
82 Loc cit.
83 Loc cit.
84 Burke 1775, p122.
85 Addison 1980, p114.
86 Third Report form the (British) Parliamentary Report on Turnpikes and Highways. 1809
87 Phillips 1737, pL4.
88 Webb & Webb 1913, p72.
89 Jackman 1916, p253.
90 Bourne 1763.
91 Mather 1696, Section 111.
92 Copeland 1968, p30.
93 Bramley 1805, p1.
94 Parsons 1939, p303.
95 Young 1809, p324.
96 Lay 2018, Chapter 5a.
97 Albert 1972, p236.
98 Albert 1972, p136.
99 Copeland 1868, p35.
100 Bird 1969, p47.
101 Smiles 1874, pp1–73.

CHAPTER 4

102 ASTM Standard D2487, Standard Practice for Classification of Soils for Engineering Purposes (Unified Soil Classification System).
103 Seely 1984.
104 Wooltorton 1954, p257.

105 Lay 2009, Figure 8.5
106 Lay 2009, Section 8.4.2.
107 Lay 2009, Section 8.4.2e.
108 Porter 1938, pp330–2 & Lay 2009, Section 8.4.2.
109 Ingles & Metcalf 1972, Chapter 7.
110 Lay 2009, Section 8.3.3.
111 Rose 1953, pp91–2.
112 Aitken 1900, p94.
113 Phillips 1737, pL9.
114 Lay 1992, p72.
115 Leighton 1972.
116 An Act for the continuance of the statute made 2 & 3 P. & M. for the amendment of highways. 5 Elizabeth, Cap XIII:554 (III). 1562.
117 Lay 2009, Section 8.6.
118 Hosking 1992, pp82–5.
119 Hosking 1992, p4.
120 DKC 1878, p331.
121 Lay 2009, Section 8.6b1.
122 Lay 2009, Section 8.6b.
123 Lay 2009, Section 8.2.2.
124 Lay 2009, Section 8.4.1.
125 Talbot & Richart 1923, pp1–116.
126 Lay 1993, p18.
127 Lay 2009, Section 8.6.
128 Gillmore 1876.
129 Lay 2009, Section 8.6r.
130 Lay 2009, Section 14.3.3.
131 Jameson 1996, p11 & Austroads 2014.
132 Lay 1993, p26.

CHAPTER 5

133 Wikipedia.
134 Newcomen 1951 & Schreiber 1961, p6.
135 e.g. Proctor 1607.
136 Schreiber 1961, p162.
137 Photo courtesy of Victorian Department of Transport.
138 Busch 2008, pp244–5.
139 Jordan 1985, p305
140 Forbes 1934, p39.
141 Forbes 1934, pp39–44 & Hindley 1971, p11.
142 Photo with kind permission of Jim Brandenburg, Minnesota.
143 Coles 1984, Coles &Coles 1986 & Ellis 2003.
144 Coles & Coles 1986, Figure 6.
145 Coles & Coles 1986, p45.
146 Taylor 1979, p13.

147 Coles & Coles 1986.
148 Forbes 1934, p45.
149 Coles & Coles 1986, Plate 34. Photo copyright Somerset Levels Project and used with its kind permission.
150 Aitken 1900, p5, Hindley 1971, p11 & Chevallier 1976, pp89–91.
151 1955 drawing by J Mertens in Chevallier 1976, p90.
152 MacAlister 1932.
153 Gillespie 1847, p230.
154 Rose 1953, pp70–1.
155 Baker 1903, p274.
156 Federal Highway Administration 1976, p51.
157 Sometimes called "furze".
158 Deveraux, R 1936, p4.
159 Phillips 1737, pL5.
160 Albert 1972, p137.
161 Lay 1997. Sometimes called Metcalfe.
162 Law 1850, pp98&109.
163 Foote 1994, 7.
164 Leahy 1844, pp87–8.
165 Fortidsnindeforvaltning undated.

CHAPTER 6

166 Bergier 1622.
167 Needham 1971.
168 Nylan 2012, p39.
169 Nylan 2012, p36.
170 Japan Road Association 2012.
171 Leighton 1972, p40.
172 Bodey 1971, Figure 5.
173 Addison 1980, p32.
174 Lay 1992, Figure 3.4.
175 Addison 1980, p31.
176 Lay 2018, Ch 5b.
177 Associated Press, 26 June 2012 (http://trib.com/news/scienec/greece-subway).
178 Davies, p60.
179 Photo by Brian Harper.
180 Addison 1980, p31.
181 Davies 2002, p82.
182 Addison 1980, p31.
183 Bergier 1622.
184 Chevallier 1966, p87.
185 Davies 2002, pp55–66.
186 French 1981, p19.
187 Chevallier 1966, p64 & Merdinger 1952, p269.

CHAPTER 7

188 Lay 1992, Chapter 4.
189 Chartres 1981, p65.
190 Parsons 1939, p291.
191 Hindley 1971, p78.
192 Hindle 1982, p13.
193 Williams 1998.
194 Lay 1992, p64.
195 Calendar of Letter Books of the City of London, Malcolm 1933, p85.
196 Moaligou 1982, p39 & Parsons 1939, pp231–4.
197 From C Weigel's self-published Book of Estates via Schreiber 1961, p16.
198 Collins & Hart 1936, p3.
199 Parsons 1939, pp265–7.
200 Jusserand 1901, pp84–6 & 44.
201 Jusserand 1901, pp37–41.
202 Pawson 1977, p68.
203 Jackson 1980, p65.
204 Malcolm 1933, p87.
205 Parsons 1939, p231, p239.
206 Lay 1992, Chapter 4.
207 Busch 2008. p242.
208 Busch 2008, p245.
209 Act 1562.
210 O'Keeffe 1973, p54.
211 O'Keeffe 1973, p54.
212 Chevallier 1966, p63.
213 McCloskey 1949, p353.
214 Boutet 1947.
215 Corcoran 2006, p795.
216 Leahy 1844, p14.
217 Dunham 1955, pp16–24.
218 Jackman 1916.
219 Deveraux 1936, p58.
220 Clark 1794.
221 Newlon, Pawlett et al. 1985, p35.
222 Anon 1980, pp16–8.
223 Lay 1992, p102.
224 Federal Highway Administration 1976, p37.
225 Act 1836.
226 Leahy 1844, pp35–62.
227 Mullins 1859, p109.
228 O'Keeffe 1973, p63.
229 Kautilya c300 BC.
230 Leighton 1972.
231 Parsons 1939, p293.
232 Lay 2018, Chap 11d.
233 Guldi 2012, p209.

CHAPTER 8

234   Parsons 1939, p298.
235   Moch 1947.
236   Toglietta 1585 & Parsons 1939, pp292–3.
237   Bergier 1622.
238   Parsons 1939, p292.
239   Lay 1992, Figure 3.8a.
240   Aitken 1900, p5.
241   Gautier 1693.
242   Lay 1992, Figure 3.8
243   Lay 1992, Figure 3.8
244   Lay 1992, pp112–4.
245   Trésaguet 1831.
246   Lay 1992, Figure 3.8.
247   Dunham 1955, p16.
248   Borth 1969, p65.
249   Boulnois 1919, p85.
250   Lay 2018, Chapter 4A.
251   Defoe 1725, p130.
252   Defoe 1725, pp117–8.
253   Webb & Webb 1913, p75.
254   Albert 1972, Chapter 7.
255   Pawson 1977, p241.
256   Rolt 1958.
257   Haldane 1962.
258   Bird 1969, p52.
259   Haldane 1962, p68.
260   Southey 1829.
261   Aitken 1900, p250.
262   Rickman 1838, Figs 1 & 2.
263   Aitken 1900, p250.
264   Aitken 1900, pp15–18.
265   Cresy 1847.
266   Aitken 1900, p19.
267   Parnell 1838 p70.
268   Lay 1992, Figure 3.8.
269   Aitken 1900, Figure 116.
270   Aitken 1900, p16.
271   Lay 2018, Chapter 4C.

CHAPTER 9

272   Lay 2018, Chapter 4C.
273   Lay 2018, Chapter 1E.
274   Lay 1992, p75.
275   Lay 2018, Chapter 4A.

276 Manton 1956, p6.
277 1825 engraving by Charles Turner, from British Museum collection.
278 Addison 1980, p117.
279 Aitken 1900, p14.
280 Lay 1992, Figure 3.8e.
281 McAdam 1816, p40 in 1824 8th edition.
282 Deveraux 1936, p50.
283 Lay 2009, Section 8.3.2.
284 Albert 1972, p158.
285 Aitken 1900, p23.
286 Kellett 1969, p345.
287 Law 1850, p119.
288 Holley 2003, pp703–33, Aitken 1900, pp81–5 & Rose 1953, pp77–8.
289 Baker 1903, Figure 59.
290 Lochhead 1878.
291 Lay 1992, p78.
292 Phillips 1737, pL10.
293 McAdam 1824, p106.
294 Lay 2009, Section 11.1.3.
295 Merdinger1952, pp268–73.
296 McAdam 1824, p51.
297 McAdam 1824, p41.
298 McAdam 1824, p41.
299 McAdam 1819, p22.
300 McAdam 1824, p40.
301 Aitken 1900, p25e.
302 Lay 2018, Chapter 11.
303 Young 1809, p324.
304 Forbes 1953.
305 Stevenson 1824.
306 Leahy 1844, pp15–6.
307 Baker 1903, p210.
308 McAdam 1816.
309 McAdam 1819.
310 Department of Main Roads (NSW) 1976, p36.
311 Rose 1953, pp52–3.
312 Lay 2018, p96.
313 Reverdy 1980, pp39–43; Aitken 1900, p6 & Dunham 1955, p20.
314 Codrington 1892.
315 North 1879, p98.
316 North 1879, p103.
317 North 1879, p112.
318 Law 1850.
319 Needham 1971.
320 McShane 1979, p297.
321 Japan Road Association 2012.
322 Parnell 1838, pp24–5.
323 Albert 1972. e.g. p143 & Parnell 1838, 24–5, 78–9, 124–9.

324 Paterson 1819.
325 Jackman 1916, p276.
326 McAdam 1816 & 1824, p50.
327 Moore 1876.
328 Lochhead 1878.
329 Clarke 1965, p220.
330 Edgeworth 1817, p19.
331 Anderson 1932, p188.
332 Lay 1992, p49.
333 Pierre & Cohrs 1998, Figure 25.
334 Illustrated London News, 26 Apr 1851 and Mary Evans Picture Gallery.
335 Leupold 1725.
336 Burgoyne c1860.
337 Law and Clark 1907, p431.
338 Aitken 1900, p215.
339 Merdinger 1952, p273.
340 Burgoyne 1843 & Lay 2018, p306.
341 Boulnois 1919, p73 & Plate V.
342 McAdam 1824.
343 Collins & Hart 1936, p29.
344 Law & Clark 1901, p147.
345 Baker 1903, p222.
346 North 1879, p100.
347 Lay 2018, Chapter X.
348 Aitken 1900, p216.
349 Anon 1876a, p298.
350 Pierre & Cohrs 1998, Figure 26.
351 Deacon 1879.
352 Lay 1992, p85 & Law and Clark 1907, p434.
353 Kennerell 1958, p189.
354 Aveling-Barford Company 1965.
355 Lay 2018, Ch 14.
356 Aveling-Barford, 1965, p11.
357 Crosbie-Dawson 1876, p20.
358 Boulnois 1919, p73 & Plate V.
359 McAdam 1824, pp46–7.
360 McAdam 1825, p24.
361 McAdam 1824, p50.
362 Forbes 1958a, p534.
363 Hartman 1927.
364 Law 1850, p97 & McAdam's evidence to the 1819 Parliamentary Committee recorded on p111 of McAdam 1824.
365 Lay 2018, Chapter 16.
366 Judson 1908, p11.
367 Hadfield 1934, p83.
368 Lay 2018, Chapter 16b.
369 Sachs 1992, p14.
370 Sauterey 1997, p77.

371 Hubbard 1910, p2.
372 Boulnois 1919, p99.
373 Byrne 1917, p164.
374 Hoffbeck 1991.
375 Anon 1895.
376 Hveem 1971, p24.
377 Spencer1965, p33 & Judson 1908.
378 Hubbard 1910, p1.

CHAPTER 10

379 Boulnois 1919, p5.
380 Lay 1992, p204.
381 Lay 1992, p204.
382 Merdinger 1952, p340 & Moaligou 1982, p39.
383 Rose 1952, p23.
384 Bird 1969, p48.
385 Hyslop 1984.
386 Collingwood 1891.
387 Aitken, 1900, p312.
388 Baker 1903, p525.
389 Photograph by the authors.
390 Parsons 1939, p231.
391 Mountain 1894, p8.
392 Aitken 1900, p304 & Baker 1903, p539.
393 Hugo 1862, p988.
394 Moore 1876.
395 Crosbie-Dawson 1876, p23.
396 Collins & Hart 1936, p33.
397 Aitken 1900, p313.
398 Franklin 1843, pp163–8.
399 Photo by the authors.
400 Deacon 1879, p2.
401 Fortune 1893, p902 & 910.
402 Merdinger 1952, p340.
403 Franklin 1843, p150.
404 James 1964a.
405 Abraham 1960, p49.
406 Baker 1903, p553 & McShane 1994, p59.
407 Gross 1901, pp206–7.
408 Franklin 1843, p151.
409 Collingwood 1891.
410 Dickens 1856, p130.
411 Moore 1876, p163.
412 St George 1879.
413 Mountain 1880, p29.
414 McShane 1994, p60.

415 Lay 1992, Table 7.2.
416 Haywood 1871.
417 Delano 1880.
418 Mountain 1903.
419 Report 1886.
420 Aitken 1900, pp323–7.
421 Mountain 1887.
422 Aitken 1900, pp331–42.
423 Boulnois 1919, p227.
424 Mountain 1913
425 Crosbie-Dawson 1876, p51.
426 ANAS 2003, p48.photo courtesy of Professor Enrico Menduni
427 Reilly 1918.
428 Moaligou 1982, p41.
429 Aitken 1900, p359.
430 Baker 1903. p462.
431 Sharp & Armstrong 1985.
432 Lay 1992, p229.
433 Sharp & Armstrong 1985 (Author's photo).
434 Lay 2018, p134.
435 Crosbie-Dawson 1876, p25.
436 Hadfield 1934, p211.
437 Merdinger 1952, p341.
438 Moaligou 1982, p41.
439 Anon 1930, p53.
440 Anon 1856, Law & Clark 1907, p321 & Crosbie-Dawson 1876, p51.
441 Anon 1943.
442 Moaligou 1982, p41.
443 Hadfield 1934, p206.
444 Law & Clark 1907.

CHAPTER II

445 Lay 2009, Section 8.7.1.
446 Lay 2009, Sections 8.7–9.
447 Dickinson 1984, p36.
448 Lay 2009, Section 8.7 & Dickinson 1984, p224.
449 Hveem 1971, p29.
450 Dickinson 1984, p50 & Lay 2009, Section 8.7.7.
451 Sharp, Patrick, Thananjeyan & Simpson, 2012.
452 Forbes 1958b, p102.
453 Deacon 1879, p6.
454 Halstead and Welborn 1974, p94.
455 Lay 1992, p244 & Halstead & Welborn 1974, pp113–4.
456 Boulnois 1919, p187.
457 Chabrier 1876, p286.

458   Delano 1893, Figure 29.
459   Lay 2009, Section 8.8.
460   Merdinger 1952, p342.
461   Lay 2018, p234.
462   Hughes, Adam & China 1938, p169.
463   Lay 2009. Section 3.6.2.
464   Lay 2009, Section 8.9.
465   Lay 2009, Section 8.5.4.
466   Lay 2018, Chapter 15a.

CHAPTER 12

467   Anon 1795.
468   Attributed to J R Smith in Anon 1795 or to an engraving by P. Rayner.
469   Lay 1997 & Anon 1795.
470   Manton 1956.
471   Reader 1980, p36.
472   Mountain 1903.
473   Lay 1984, p18.
474   Anon 1912, p6.
475   City of Sydney Archives.
476   Mountain 1903, p4.
477   Mountain 1885.
478   Mountain 1903, pp69–70.
479   Gillmore 1876, p205.
480   Mountain 1903, p74.
481   Hadfield 1934, p116
482   Lay 1992, p204.
483   Mayhew 1850, p369.
484   photo©The Heath Robinson Museum.
485   Parsons 1939, p229, 232, 267–8 & Lay 1992, p205.
486   Lay 2010, p64 & Parsons 1939, pp265–7, 272.
487   Mayhew 1850, p227.
488   https://en.wikipedia.org/wiki/ General_Order_No._28.
489   The Holy Bible, Apocrypha, Daniel 3:23.
490   The Holy Bible, Revelations, 21:21.

CHAPTER 13

491   Source and translation from Bilkadi 1996, p138.
492   James 1964a.
493   Aitken 1900, p367.
494   Abraham 1960, p47.
495   BC 1824.

496 Boutet 1947, p14.
497 Earle 1974, p8.
498 Mountain 1887.
499 Richardson 1908, p375.
500 Warren 1901.
501 Boulnois 1919, Chapter 5.
502 Ritchie 1999, p10.
503 Earle 1971.
504 Rose 1953, p105.
505 Lay 1992, pp216–7.
506 Ritchie 1999, p42.
507 Lay 2009, Section 3.6.2.
508 Lay 1992, pp216–7.
509 d'Eyrinis 1721, p158.
510 Ironbridge Gorge Museum Trust, 1979.
511 Lay 1992, pp216–7.
512 Meyn 1873.
513 d'Eyrinis 1721, p158.
514 Lay 1992, p50, 242–3.
515 A detailed English-language story of these and other asphalt deposits is available in Part 1 of Spielmann & Hughes (1936).
516 Delano 1893, Figure 2.
517 Abraham 1960, pp171–86.
518 Broome 1963 & d'Eyrinis 1721.
519 Lay 1992, p208, Meyn 1873, p77 & Malo 1866.
520 Broome 1963, p1438.
521 Forbes 1958a & Meyn 1873.
522 Meyn 1873, p78.
523 Abraham 1960, p46, Meyn 1873 & Lay 1992, p209.
524 Abraham 1960, p47.
525 Meyn 1873, p79.
526 Abraham 1960, p47.
527 James 1964a.
528 Franklin 1843, p148.
529 Franklin 1843.
530 Ellice-Clark 1880.
531 Contemporary sketch by George Scharf. With the permission of the Hulton Picture Company and from their Hulton-Deutsch Collection via Getty Images.
532 Darcy 1850, p186.
533 Gillmore 1876, p191.
534 Broome 1963, p1472.
535 Meyn 1873, p79.
536 de Coulaine 1850, Chapter 6.
537 Broome 1963, p1437.
538 Malo 1866 (Broome 1963 p1437 translation).
539 Lay 1992, p216.
540 Lay 1992, p213.
541 Broome 1963, p1437.

542   Malo 1866.
543   Chabrier 1876, p284.
544   Chabrier 1876, p279.
545   Ellice-Clark 1880, p155.
546   Law 1850, p99.
547   Meyn 1873, p124.
548   Ellice-Clark 1880, p155.
549   Lay 2018, Ch 11e.
550   Meyn 1873, p79 & 124.
551   de Coulaine 1850, Chap 1.
552   Darcy 1850, p186, Malo 1866, pp196–9.
553   Chabrier 1876, p281.
554   Holley 2003, p707.
555   Peckham 1909, p288.
556   Lay 2018, p226.
557   Boutet 1947, p14.
558   Aveling-Barford 1965, p6.
559   de Coulaine 1850.
560   Abraham 1960, p49.
561   Lay 1992, p215.
562   Anon 1876b.
563   Delano 1893.
564   Malo 1886, p7
565   Collingwood 1891.
566   Broome 1963, p1438.
567   Anon 1890.
568   Mountain 1903, pp39, 44 & 74
569   Broome 1963, p1439.
570   Crosbie-Dawson 1876, p39.
571   Meyn 1873, p124.
572   Aitken 1900, Chapter 11.
573   Aitken 1900, Figure 126.
574   Holley 2003, p707.
575   Krchma & Gagle 1974, p26.
576   Tillson 1897.
577   Lay 2009, Section 8.3.2 & Forbes 1958a, p540.
578   Warren 1901, p199.
579   Lay 1992, p232.
580   Griffin 1887.
581   Robertson 1958, p6.
582   Gillespie 1992 & Anon 1891.
583   Anon 1891.
584   Boulnois 1919, p181.
585   Richardson 1908, p3.
586   McNichol 2005, p64.
587   McShane 1979, p295.
588   Greene 1885, p186.
589   Armstrong 1976.

590 McShane 1994, Graph 4.1.
591 Tillson 1897.
592 Federal Highway Administration 1976.
593 https://www.fhwa.dot.gov/highwayhistory/dodge.
594 Office of Public Roads 1913.
595 Forbes 1958a, p540.
596 Deacon 1879, p15.
597 Gillmore 1876, p203.
598 Greene 1885, p186.
599 Malo 1886, p6, 19.
600 Boulnois 1919, Tilson 1901, Deacon 1879, Agg 1916 & Byrne 1917.
601 Baker 1903, p583.
602 Morin 1842.
603 Lay 2018.
604 Carey 1914, p10 & Baker 1903, pp21–30.

CHAPTER 14

605 Agg 1916, Figure 98.
606 Deacon 1879, p10.
607 Aitken 1900, p367.
608 Lay 1992, p246.
609 Lay 1992, p246.
610 Halstead & Welborn 1974, p115.
611 Bird 1969, p61.
612 Gregory 1931, p253.
613 Anon 1989, p89
614 Distin 1992.
615 Hadfield 1934, p5 & 82.
616 Hadfield 1934, facing p643.
617 Hadfield 1934, p96.
618 Lay 1992, p246–7.
619 Lay 2009, Section 12.1.5a.
620 Hanson 1935.
621 Towler & Dawson 2008.
622 Road Research Laboratory 1962, Chap. 24.
623 Lay 2009, Section 8.7.8.

CHAPTER 15

624 Holley 2003, p718.
625 Green & Ridley, p38.
626 Broome 1963, p1475.
627 Spencer 1965, p48.
628 Spencer 1965, p49.

629   Judson 1908, p100.
630   Lay 1992, p239 & Broome 1963, p1475.
631   Tunnicliff et al. 1974, p160.
632   Richardson 1908, p378.
633   Holley 2003, p718.
634   Tunnicliff et al. 1974, p166.
635   Crawford 1989, p13.
636   Tunnicliff et al. 1974, p170, 212.
637   Early versions are shown in Agg 1940, Figure 103.
638   Broome 1963, p1475.
639   Tunnicliff et al. 1974. p164.
640   Richardson 1905.
641   Halstead & Welborn 1974.
642   Richardson 1908.
643   Hveem 1971, p26.
644   Crawford 1989, p13.
645   Hveem 1971, p27 & Monismith and Brown 1999, p140.
646   Anon 1890.
647   Broome 1963, p1475.
648   Lay 2009, Section 12.5.3.
649   Austroads 2019a, Appendix A.1.18, p177.
650   Broome 1963.
651   Earle 1974, pxvi.
652   Law 1962.
653   Road Research Laboratory 1962.
654   Hadfield 1934, p82.
655   Hadfield 1934, p120 & Speilmann & Hughes 1936.
656   Gillespie 1992, p43 & Hveem 1971, p23.
657   Holley 2003, p720.
658   Crawford 1989, p10.
659   Weingroff 2008.
660   Holley 2003, p732.
661   Lay 2009, Section 12.2.2(f).
662   Lay 2009, Section 12.2.2.
663   Gillespie 1992, p50.
664   McNicol, 2005, p157, 254.
665   Gillespie 1992, Chapters 4-7.
666   Lay 2009, Sections 8.7.3 & 8.7.8.
667   Hadfield 1934, p92.
668   Photo and data kindly supplied by Wirtgen.
669   Delano 1880, p261.

CHAPTER 16

670   Lay 2009.
671   Porter 1949, p11.

672   Porter 1938, p325.
673   Porter 1949, Figure 1, With permission from ASCE.
674   Porter 1949, p16.
675   McNicol 2005, pp178–80.
676   Jameson 1996, p4, 5.
677   Attwooll 1955, p73.
678   Porter 1949, Figure 5, with permission from ASCE.
679   Gillespie 1992, p118.
680   Monismith & Brown 1999, p129.
681   Lay 2009, Section 11.4.
682   Lay 2009, Section 27.3.4.
683   Ioannides & Khazanovich 1993.
684   Transportation Research Board 2007.
685   Gillespie 1992 p44.
686   AASHO 1961.
687   Lay 2009, Figure 11.13.
688   Monismith & Brown 1999, p152.
689   Lay 2009, Section 11.2.3.
690   Shell 1978.
691   Powell et al. 1984, p1.
692   Monismith & Witczak 1983.
693   Vuong 2008.
694   Photo by Con Sinadinos of Pavetest.
695   Jameson 1996, p1, 8.
696   Hveem & Carmany 1948, p104.
697   Seely 1984.
698   Terzaghi 1960.
699   Kane undated.
700   Austroads 2004, p8.19, Example Chart EC13.
701   Halladay 1998.
702   Harman et al. 2002.
703   Holley 2003, p731.
704   Lay 2009, Chapters 8–14.

CHAPTER 17

705   Lay 2018, Chapter 4c.
706   Lay 2009, p220.
707   Law 1850, p101, 111.
708   Aitken 1900, p309.
709   Deacon 1879, p9.
710   Peckham 1909, p285.
711   Borth 1969, p214.
712   Ray 1964.
713   Lay 2018, p88.
714   Boulnois 1919, Plate XIII & Deacon 1879, p10.

715 Lay 1992, p221.
716 Baker 1903, p272.
717 Rose 1953, p106.
718 Boulnois 1919, p258.
719 Boulnois 1919, pp251–63.
720 http://overlays.acpa.org/Concrete_Pavement/About_Concrete/Evolution_of_ Design/index.asp.
721 PIARC 1923.
722 Ray 1964.
723 Ray 1964, p82.
724 Mountain 1927, p436.
725 Boussinesq 1885 & Lay 2009, Section 11.2.3.
726 Westergaard 1926.
727 Lay 2009, Section11.3.1.
728 Collins & Hart 1936, pp499–501.
729 Burmister 1943.
730 Monismith & Brown 1999, p151.
731 Portland Cement Association 1984.
732 Anon 1937.
733 Vahrenkamp 2010, p15.
734 McNicol 2005, p214.
735 www.wirtgen.de/en/products/slipform-paver.
736 www.fhwa.dot.gov/pavement/concrete/pubs/hif16003.pdf.

CHAPTER 18

737 Catton 1959.
738 Lay 1992, p223.
739 Based on Austroads 2019b.
740 Austroads 2019b.
741 www.wirtgen.de.
742 Austroads 2019b.
743 Lay 1996.
744 Agg 1940, p369.
745 Agg 1940, Chap. IX.
746 Revue Générale des Routes et des Aérodromes 1976, p74, Fig 3.
747 Baker 1903, p271.
748 Crawford 1938.
749 JB Metcalf photo.

CHAPTER 19

750 Lay 2009.
751 Macneill 1833.

752 Coane, Coane & Coane 1908+, pp178–83 in 1927 edition & Markwick & Shergold 1945.
753 Lay 1992, pp37–41.
754 Morin 1842.
755 Lay 2018, p7.
756 Photo courtesy of Victorian Department of Transport.
757 Federal Highway Administration 1976. Ch 9, pp90–3.
758 Lay 1992, p172.
759 Lay 2009, pp205–6.
760 Anon 1952.
761 Lay 1993, p23, Figure 5.
762 Auff 1982.
763 Hveem 1955.
764 Lay 2009, Section 14.5.2.
765 Croney 1997, p331.
766 The authors & Anon 1955.
767 Lay 2009, Section 14.5.2.
768 Lay 2009, Section 11.3.4.
769 www.dynatest.com.
770 Descornet 1990.
771 Boulnois 1919, p289.
772 Giles & Sabey 1959.
773 Lay 2009, Section 8.6(o).
774 Hosking 1992, p160.
775 Courtesy of WDM, who has been the sole licensed manufacturer worldwide of the SCRIM vehicle since the early 1970s.
776 James 1964b.
777 Martin 2005.
778 Lay 2009, Section 14.4.
779 Boulnois 1919, p202.
780 Coffee 1900.
781 Aitken 1900, p420.
782 Hogentogler 1923.
783 Hveem 1960, p3.
784 Hveem 1960, pp16–20.
785 Hveem 1960, p11.
786 Hogentogler 1923.
787 ERPUG 2015 (includes visual review).
788 www.floridamemory.com/items/show /105566. Photo with kind permission of State Archives of Florida.
789 NPL records, See also Croney 1997, Figure 26.1.
790 Hveem 1960, pp8–11.
791 Hogentogler 1923.
792 Anon 1926.
793 Hveem 1960, p4.
794 Photo courtesy of Victorian Department of Transport.
795 Buchanan & Catudal 1941.
796 Kier 1974.

797　Scala 1970.
798　Anon 1926, p148.
799　Hudson & Hain 1961 & Huft 2010.
800　Sayers 1995, p1.
801　Reprinted with permission from Prem 1989, Figure 5, arrb.com.au.
802　Carey & Irick 1962 & Carey, Huckins & Leather 1973.
803　Reprinted with permission from ARRB, arrb.com.au.
804　Reprinted with permission from ARRB, arrb.com.au.
805　Lay 2009, Section 14.4.3
806　Reprinted with permission from "ARRB the first 50 years" (Sharp et al. 2011), arrb.com.au.
807　Gillespie, Paterson & Sayers 1986.
808　Sayers, Gillespie & Queiroz 1986.
809　Watanatada et al. 1987.
810　Based on Sayers & Karamihas 1998.
811　Martin 1996.
812　Martin 1996, p34.
813　Gillespie 1847, p147.
814　Chamberlin 1995, p7.
815　Auff 1982.
816　Chamberlin 1995.
817　Lay 2001.
818　Haas, Hudson & Cowe-Falls 2005, p5.
819　Lay 2009, Figure 14.1.
820　Lay 1999.

CHAPTER 20

821　Metcalf 1996 & 2014.
822　Crompton 1913 & 1928.
823　Croneys 1997.
824　Croneys 1997.
825　Croneys 1997, p5.
826　Transport and Road Research Laboratory 1960.
827　Powell et al. 1984.
828　Powell et al. 1984, p2.
829　Clarke 1965.
830　Kirwan 1911.
831　Anon 1913.
832　Hines 1913, Anon 1913.
833　Hines 1913.
834　Anon 1913.
835　Emmons 1934.
836　Anon 1918–1924.
837　FHWA 1976, p332.
838　Thomson 1928 & FHA 1976, p331.

839 Crompton 1913.
840 Boulnois 1919; Speilmann & Hughes 1936.
841 Pyatt 1983.
842 Public Roads, May 1924.
843 Photo courtesy of National Physical Laboratory.
844 NPL Annual Report 1911.
845 NPL Annual Report 1911–1933.
846 Source unknown.
847 Historic concrete pavements @ explorer.acpa.org/explorer.
848 Mentioned by Ray 1964 but photo source know.
849 Metcalf 1996.
850 OECD 1985, p22.
851 Ravensteijn & Kop 2008.
852 Berthier 2013.
853 Anon 1933.
854 Carpenter & Willis 1938.
855 Metcalf 2014.
856 Weingroff 2008.
857 Anon 1954.
858 Carpenter & Goode 1936.
859 Metcalf 2014.
860 Morris 1972.
861 Anon 1955 & 1962 & Croney 1997, p268.
862 Ekse & LaCross 1957.
863 Sparks et al. 1968 & 1970.
864 JB Metcalf photo.
865 JB Metcalf photo.
866 Croneys 1997, p252.
867 Mahoney& Terrel 1982.
868 Ekse & LaCross 1957.
869 Terrel & Krukar 1969.
870 From http://explorer.acpa.org/explorer/ and from the cover of the Summary Report, Highway Research Board Special Report 61G, 1961.
871 du Plessis et al., 2008, p17–33.
872 JB Metcalf photo.
873 https://www.csir.co.za/heavy-vehicle-simulator.
874 Ullidtz 1982 & Dohmen & Molenaar 1992.
875 https://sites.google.com/site/afd40web/apt-facilities-outside-the-usa/europe. Permission to use image from Stefan A. Romanoschi.
876 https://sites.google.com/site/afd40web/apt-facilities-outside-the-usa/europe. Permission to use image from Stefan A. Romanoschi.
877 OECD 1985, p22.
878 Berthier 2013.
879 Kerzrého 2013.
880 JB Metcalf photo.
881 Rose & Bennett 1994.
882 J Johnson-Clarke photos. Reprinted with permission from ARRB, arrb.com.au.

883   Behr 1977.
884   JB Metcalf photo.
885   Shackel & Arora 1978.
886   JB Metcalf photo.
887   Metcalf 2014, pp32–6.
888   Sharp 2004.
889   Metcalf 1996.
890   Hugo & Epps-Martin 2004.

CHAPTER 21

891   Grübler 1990, Figure 3.3.1 (p87).
892   Photo by Richard Yeo.
893   www.mdpi.com/1424-8220/17/9/2053.
894   McNichol 2005, p12 & 262.
895   www.fhwa.dot.gov/publications/research/infrastructure/structures/97148/ wg2.cfm.
896   Lay 2018, p211.
897   www.pinehurstseattle.org/2008/08/15/pretty-bioswales.

CHAPTER 22

898   Lay 2009, Chapters 29–31.
899   Used with the kind permission of The Public Interest Incorporated Association, Aomori Tourism Federation.

Index

Printed in the United States
by Baker & Taylor Publisher Services

Printed in the United States
by Baker & Taylor Publisher Services